T0222885

Coordination Chemistry

Birgit Weber

Coordination Chemistry

Basics and Current Trends

 Springer Spektrum

Birgit Weber
Anorganische Chemie IV
Universität Bayreuth
Bayreuth, Germany

ISBN 978-3-662-66440-7   ISBN 978-3-662-66441-4   (eBook)
https://doi.org/10.1007/978-3-662-66441-4

This book is a translation of the original German edition "Koordinationschemie" by Weber, Birgit, published by Springer-Verlag GmbH, DE in 2021. The translation was done with the help of artificial intelligence (machine translation by the service DeepL.com). A subsequent human revision was done primarily in terms of content, so that the book will read stylistically differently from a conventional translation. Springer Nature works continuously to further the development of tools for the production of books and on the related technologies to support the authors.

This Springer Spektrum imprint is published by the registered company Springer-Verlag GmbH, DE, part of Springer Nature.
The registered company address is: Heidelberger Platz 3, 14197 Berlin, Germany

Learning is like rowing against the current, when you stop you drift back.
Lǎozǐ

Notes on the First Edition

This textbook is based on lectures on coordination chemistry given by me at the University of Bayreuth. The aim was to write a basic textbook with a simple and easy to understand introduction to coordination chemistry. It should fill the gap between the books on coordination chemistry for advanced students, and the short but not so comprehensive introductions to coordination chemistry from textbooks on general and inorganic chemistry. This textbook is intended for undergraduate chemistry students, but also for subsidiary subject students such as student teachers, biologists, biochemists, and anyone else who wants to expand their knowledge of coordination chemistry. For this reason, the book is designed to be self-contained. The last three chapters provide brief insights into current trends and research directions.

Notes on the Second Edition

In the second revised edition of this textbook, errors and inaccuracies gathered within the last 7 years have been weeded out. In addition, two sections have been added to the book. In the new Chap. 11, the fundamentals of luminescence of coordination compounds are laid out. In addition, a section on photocatalysis has been added to the catalysis chapter. Both of these newly added sections deal with very current areas of research and are aimed at already advanced students, but should be easy to understand in conjunction with the other chapters of the book.

Acknowledgements

First of all, I would like to thank all the readers of the first edition of this book for the consistently positive feedback that reached me. Without this feedback, I would not have been so motivated to work on a second edition. Special thanks go to all those students who pointed out errors and inconsistencies in the book to me, which have been corrected in this new edition. I would like to thank, in alphabetical order: J. Breternitz, S. Freund, F. Gruschwitz, J. Lipp, C. Memmel, A. Methfessl, N. Müller, J. Petry, T. Rößler, O. Scharold, J. Simon, K. Soliga, H. von Wedel, P. Weiss.

I would like to thank my co-workers, especially Hannah Kurz, Sophie Schönfeld and Andreas Dürrmann, as well as Dana Dopheide, for being available as proofreaders for the whole script and for accepting without complaint that I put time into this project (and left other things undone).

I would like to thank Prof. Dr. Katja Heinze, Prof. Dr. Roland Marschall, Prof. Dr. Peter Klüfers and Dr. Gerald Hörner for extensive technical advice, especially on the new sections on luminescence and photocatalysis. For help with the proof corrections of the English edition, I thank Dr. Gerald Hörner, Andreas Dürrmann, Florian Daumann, Constantin Schreck, Sebastian Egner, and Dana Dopheide.

As with the first edition, my very special thanks go to my family, my husband Jan and my three children Carl, Emma and Lisa. It still applies: Life wouldn't be half as good without you! Thank you for being here!

Contents

What Are Complexes?

We have contact with complexes or coordination compounds in our daily lives. We encounter them as colors, such as Prussian blue, and they are essential for the processes of life that take place in our bodies. Complexes play a central role in various large-scale technical processes. Examples include cyanide leaching, the production of high-purity aluminum oxide using the Bayer process, or the purification of nickel using the Mond process. Different stabilities and solubilities of complexes are applied in the separation of metals, including the increasingly important rare earth metals, without which many of the high-performance electronic devices would be inconceivable. The preparation of polymers under mild conditions would be unthinkable without the use of (organometallic) complexes as catalysts. And also the qualitative and quantitative analysis as it is taught at the beginning of studies or to some extent already in schools is not possible without complexes (complexometric titration, nickel gravimetry, specific cation detection reactions) [1–3]. To get started, let us consider the following examples:

1. Why does the addition of NaOH to an Al^{3+} solution first yield a precipitate that dissolves again on further addition? (Bayer process for the purification of bauxite for the preparation of aluminum).
2. Why does AgCl dissolve when NH_3 is added? (Detection of chloride ions.)
3. Why is anhydrous $CuSO_4$ colorless, an aqueous solution light blue, when Cl^- is added light green? Why does a precipitate form on the addition of NH_3 which dissolves with deep blue color on further addition of NH_3? (Fig. 1.1).

We look at the reaction equations for the processes and questions mentioned so far and the complexes that appear in them:

Fig. 1.1 Colours of copper salts and complexes. From left to right: $[CuCl_2(H_2O)_4]$, $CuSO_4$ (anhydrous), $[Cu(H_2O)_6]^{2+}$, $Cu(OH)_2$ and $[Cu(H_2O)_2(NH_3)_4]^{2+}$. The diverse colors of metal ions fascinated nineteenth-century chemists

- $Ni + 4\,CO \rightleftharpoons [Ni(CO)_4]$

$[Ni(CO)_4]$ is the complex. The colorless, readily sublimable, neutral compound allows selective separation of nickel from by-products. This Mond process, named after its discoverer Mond, is used to produce high-purity nickel on an industrial scale.

- $Al^{3+} + 3\,OH^- \rightleftharpoons Al(OH)_3\downarrow \quad Al(OH)_3 + OH^- \rightleftharpoons [Al(OH)_4]^-$

$[Al(OH)_4]^-$ is the complex. The colorless complex anion forms with Na^+ or K^+ a salt readily soluble in water (from the added base NaOH or KOH). In this way, aluminum hydroxide is separated from the non-amphoteric iron hydroxide and used to prepare highly pure alumina. An amphoteric hydroxide can react as both an acid and a base. Aluminum hydroxide is a base and can be protonated (e.g., $Al(OH)_3 + H^+ \rightleftharpoons Al(H_2O)(OH)_2^+$), but it can also react as an acid with hydroxide ions, as given in the equation above. Iron(III) hydroxide can only react as a base and is therefore not amphoteric. This difference is exploited in the Bayer process for the preparation of high-purity aluminum oxide.

- $Ag^+ + Cl^- \rightleftharpoons AgCl\downarrow \quad AgCl + 2\,NH_3 \rightleftharpoons [Ag(NH_3)_2]^+ + Cl^-$

$[Ag(NH_3)_2]^+$ is the complex. The colorless complex cation forms a salt that is readily soluble in water with the chloride ions still present in the solution. The same principle is used in cyanide leaching, but here cyanide ions, CN^-, are used as ligands, forming an even more stable complex ($[Ag(CN)_2]^-$, an anion). In this way, hardly soluble silver salts (halides, silver sulfide Ag_2S), but also elemental silver (the reaction conditions allow oxidation by atmospheric oxygen to Ag^+) as well as the other noble metals and their compounds can be brought into solution and separated from the accompanying material.

- $CuSO_4 + 6\ H_2O \rightarrow [Cu(H_2O)_6]^{2+} + SO_4^{2-}$
  $[Cu(H_2O)_6]^{2+} + 2\ Cl^- \rightleftharpoons [CuCl_2(H_2O)_4] + 2\ H_2O$
  $[Cu(H_2O)_6]^{2+} + 2\ OH^- \rightleftharpoons Cu(OH)_2\downarrow + 6\ H_2O$
  $NH_3 + H_2O \rightleftharpoons NH_4^+ + OH^-$
  $Cu(OH)_2 + 4\ NH_3 + 2\ H_2O \rightleftharpoons [Cu(H_2O)_2(NH_3)_4\ ]^{2+} + 2\ OH^-$

Complexes are $[Cu(H_2O)_6]^{2+}$ (light blue), $[CuCl_2(H_2O)_4]$ (light green) and $[Cu(H_2O)_2(NH_3)_4]^{2+}$ (deep blue). The tetraammine copper(II) ion is the first coordination compound mentioned scientifically [4].

1.1 History

The concept of complexes is best defined by its historical development [5]. Complexes have been known for over 400 years [4]. Probably the oldest compound later described as a complex is a bright red alizarin pigment, which was already used by the ancient Persians and Egyptians. It is a calcium-aluminum chelate complex with hydroxyanthraquinone, which is also used, among other things, in the cation separation pathway as an aluminum detection (detection as alizarin-S color varnish). A chelate complex has ligands that coordinate to the metal center with multiple sites. The first coordination compound scientifically mentioned is the blue tetraammine copper(II) ion, which is formed when copper ions react with ammonia. The compound was first described in 1597 by the German chemist, physician and alchemist *Andreas Libavius,* although it was not yet recognized as a complex [4]. Prussian blue (or Berlin blue or Turnbull's blue) was first produced around 1706 by the paint manufacturer *Diesbach* in Berlin. The pigment replaced the very expensive Lapis lazuli and was mentioned in several scientific letters [6, 7]. Since 1709 it was sold by dealers under the name Berlinerisch Blau or Preußisch Blau [4]. It was not until the end of the twentieth century that Turnbull's blue, produced by a different route, was shown to have the same structure as Prussian blue [8, 9]. Years later, in 1798, the French chemist *B. M. Tassaert* discovered that ammoniacal solutions of cobalt chloride turn brownish in air [4]. *Gmelin* succeeded in isolating the resulting orange complex cation in 1822 as $[Co(NH_3)_6]_2(C_2O_4)_3$. This compound belongs to a class of compounds that has been intensively studied since the middle of the nineteenth century, the ammonia adducts of cobalt(III) salts of the general composition $CoX_3 \cdot n\ NH_3$, with n = 3, 4, 5, 6. In these compounds, different (but not all!) ammonia contents could be realized, which significantly influenced the properties of the compounds. The most obvious property was coloration, which was reflected in the naming (luteo = yellow, purpureo = red, praseo = green, violeo = purple).

At that time, theories on the constitution of complex compounds were strongly oriented towards the chemistry of carbon, in which a correspondence between valence and bonding (= coordination number) was assumed. In 1870, *Blomstrand* proposed that the ammonia molecules in these compounds should be linked into chains like the CH_2 groups in hydrocarbons. This idea was inspired by the structure of diazo compounds, in which *Kekulé* had postulated a nitrogen-nitrogen bond. This so-called "chain theory" was one of the most successful and widely accepted approaches to the explanation of coordination chemistry and was further developed by *Jørgensen,* among others. His very systematic work in this field was to confirm and extend *Blomstrand's* idea. According to this approach, the structure for cobalt(III) complexes with different NH_3 contents was a picture in which the bonding of cobalt (3, conditioned by the oxidation state) was taken into account and the nitrogen was formulated as formally five-bonded [4]. Some examples are shown in Fig. 1.2. We recall that sophisticated techniques such as crystal structure analysis were not yet available in the century before last. The structure of compounds could only be deduced from their chemical reactivity and physical properties. The following observations were made for the luteo salt of cobalt, from which the questions that needed to be addressed in coordination chemistry can be derived.

If atmospheric oxygen is passed through an aqueous solution of $CoCl_2$, NH_4Cl and NH_3, a compound of the composition $CoCl_3 \cdot 6\,NH_3$, already mentioned, is formed. This compound was already a problem for chemists at the end of the century before last because of its composition. It was incomprehensible to them by what conditions the NH_3, that is, a neutral molecule, is bound in a salt. Moreover, they could not explain why cobalt could be oxidized from the divalent to the trivalent state with atmospheric oxygen – any textbook on

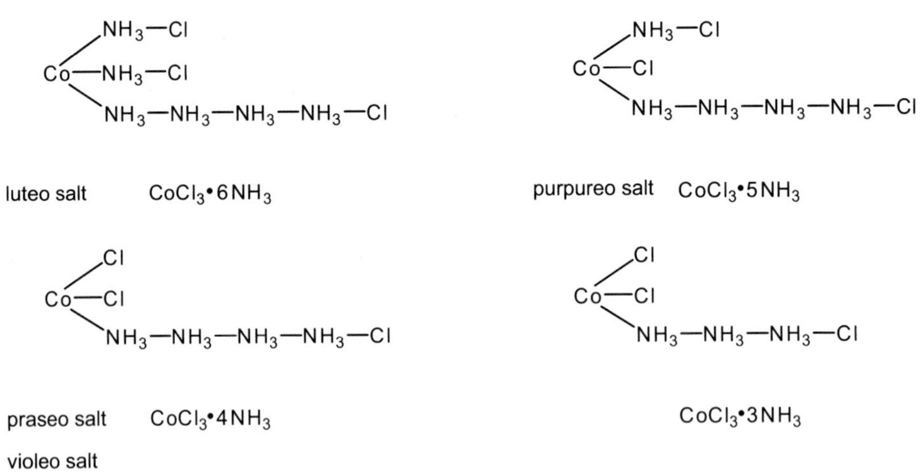

luteo salt $CoCl_3 \cdot 6\,NH_3$

purpureo salt $CoCl_3 \cdot 5\,NH_3$

praseo salt $CoCl_3 \cdot 4\,NH_3$

violeo salt

$CoCl_3 \cdot 3\,NH_3$

Fig. 1.2 Structures of cobalt(III) complexes according to *Jørgensen* based on the "chain theory". *Jørgensen* assumed that the chloride bound to the ammonia is only weakly bound and the chloride bound directly to the cobalt is very strongly bound. He was thus able to explain some, but not all, of the properties of the cobalt(III) salts with different ammonia content

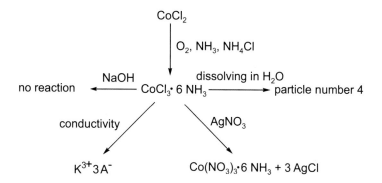

Fig. 1.3 Physical and chemical properties of $CoCl_3 \cdot 6 \, NH_3$, which were a mystery to chemists at the beginning of the nineteenth century

general and/or inorganic chemistry will tell you that the stable oxidation state for cobalt in water is +2. This compound became even more incomprehensible when its properties and reactions were examined, which are summarized in Fig. 1.3.

- Although the compound consists of ten particles, it dissociates into only four particles when dissolved in water.
- When the electrical conductivity of this solution is measured, it is found to correspond to the expected value of one trivalent cation and three monovalent anions.
- Completely unexpected is the behavior against dilute caustic soda, where precipitation of cobalt hydroxide and development of NH_3 was expected, but no reaction occurs (but only on heating).
- $AgNO_3$ solution is used to precipitate the three Cl^- ions as $AgCl$. The compound $Co(NO_3)_3 \cdot 6 \, NH_3$ is isolated from the solution.

If the amount of NH_3 and NH_4Cl is decreased in the above reaction, the other cobalt complexes already mentioned with a lower NH_3 content are obtained, namely $CoCl_3 \cdot 5 \, NH_3$, $CoCl_3 \cdot 4 \, NH_3$ and $CoCl_3 \cdot 3 \, NH_3$. Compounds with lower ammonia content could not be isolated, as well as compounds with a higher content than six NH_3 – regardless of the excess used. Furthermore, the ammonia content clearly influences the complexes' properties, e.g. the solutions' electrical conductivity in addition to the coloration already mentioned. This is shown in Fig. 1.4, conductivity decreases with decreasing NH_3 content towards $Co(NO_2)_3 \cdot 3 \, NH_3$, which is almost a non-electrolyte. For the reaction with silver nitrate, the properties are also very different. For example, for $CoCl_3 \cdot 4 \, NH_3$ only one of the chloride ions can be precipitated, for $CoCl_3 \cdot 5 \, NH_3$ there are two, and for $CoCl_3 \cdot 6 \, NH_3$ all three. The content of NH_3 molecules thus determines the type of electrolyte present and how many of the ions of the electrolyte can be precipitated, although NH_3 itself is not an ion and cannot actually participate in an ionic bond.

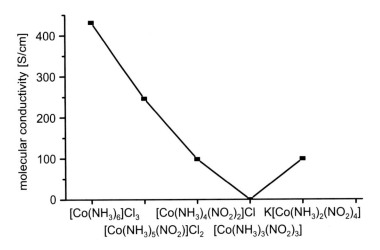

Fig. 1.4 Molecular conductivity of ammincobalt complexes in water according to [4]

Chain theory was used to explain some of the issues. The explanation was that the reactivity of the chloride ions bound via the ammonia is different from those bound directly to the cobalt. The bond to the metal should be very tight, while the interaction between ammonia and chlorine is only loose. This assumption could satisfactorily explain the properties of $CoCl_3 \cdot 6\ NH_3$, but it failed for the compound $CoCl_3 \cdot 3\ NH_3$ – it simply could not be explained why an aqueous solution of this compound showed almost no conductivity. Thus it was clear that the chain theory could only be a temporary explanation.

1.1.1   Synthesis of Cobalt Ammine Complexes

The synthesis of the cobalt ammine complexes is not trivial and exact adherence to the reaction rules is a prerequisite for success. As an example, three synthesis instructions are given below for the complexes shown in Fig. 1.5.

Luteocobalt chloride [10] 2.7 g NH_4Cl are dissolved in 6 mL water at 100 °C in a 50 mL beaker. Subsequently, 4 g $CoCl_2$ are added to the solution in portions. A spatula tip of activated carbon is placed in a 100 mL Erlenmeyer flask and the boiling $CoCl_2$ solution is added in one portion. After cooling to room temperature, 10 mL of 25% aqueous ammonia is added and the suspension is cooled to 10 °C in an ice bath. With vigorous stirring, 15 mL of 30% hydrogen peroxide solution is added. After gas evolution has subsided (1–2 h), the mixture is heated to 60°C in a water bath for 10 min. After cooling in an ice bath the resulting precipitate is filtered off, dissolved in 30 mL of 2% aqueous HCl at boiling heat and filtered hot. After the addition of 5 mL conc. HCl (37%), the solution is refrigerated. The resulting orange precipitate is filtered off and dried in air.

Fig. 1.5 Photograph of three cobalt ammine complexes. From top to bottom: Luteocobalt chloride [Co(NH$_3$)$_6$]Cl$_3$, orange; purpureocobalt chloride [CoCl(NH$_3$)$_5$]Cl$_2$, purple; and praseocobalt chloride *trans*-[CoCl$_2$(en)$_2$]Cl, green, with en instead of NH$_3$

Purpureocobalt chloride In a 125 mL Erlenmeyer flask, 1 g NH$_4$ Cl is dissolved in 9 mL concentrated ammonia solution (25%). Subsequently, 2 g CoCl$_2$· 6 H$_2$ O are added in portions, followed by 2 mL 30% hydrogen peroxide solution. The resulting solution is slowly heated to 70°C and 6 mL of concentrated hydrochloric acid (37%) is added dropwise. The solution is now stirred at 70°C for 15 min and subsequently cooled to room temperature. The resulting violet precipitate is filtered off, washed three times each with ice water and 6 M ice-cold HCl, and dried at 80°C.

Praseocobalt chloride (with ethylenediamine = en, instead of NH$_3$) [10] Dissolve 1.67 g CoCl$_2$· 6 H$_2$O in 5 mL distilled water in an ice bath with stirring. Subsequently, in the following order 6 g of a 10% aqueous ethylenediamine solution, 1.67 mL of 30% hydrogen peroxide solution and 3.5 mL of concentrated hydrochloric acid (37%) are added. The solution is now evaporated on the water bath until green crystals form at the edge of the solution. The solution is kept in the refrigerator overnight to complete crystallization. The green precipitate of *trans-CoCl$_3$*· 2 en · HCl is filtered off, washed twice with a little ethanol and three times with diethyl ether and dried at 110°C (HCl liberation).

Violeocobalt chloride (with ethylenediamine = en, instead of NH$_3$) [10] The corresponding *cis*- complex is formed from the neutral aqueous solution of the *trans*-product during heating. Any unconverted *trans*- product of a solution evaporated on the water bath can be washed out with a little cold water.

1.1.2 Complexes According to Werner: An Ingenious Impertinence

The beginning of coordination chemistry as we know it today is closely associated with the name of *Alfred Werner*. His coordination theory [11], published in 1893, was subsequently described by colleagues as "an ingenious impertinence" [5], for which he was awarded the Nobel Prize in Chemistry in 1913. Remarkably, *Werner* proposed his theory at a time when he had not yet carried out a single experiment himself. As a private lecturer at the Swiss Federal Polytechnic in Zurich, he had already been working intensively for some time on the open questions of inorganic chemistry. Tradition has it that one night he was abruptly roused from sleep with the solution of the problem before his eyes. Forcibly keeping himself awake with strong coffee, he wrote down his thoughts until the following afternoon in an essay that established the coordination chemistry of today. He spent his scientific life's work putting the theory on a firm experimental footing. The great breakthrough in *Werner's* coordination theory was the abandonment of the constraint oxidation number = bonding, in that he introduced for complexes, in addition to this "primary valence" (= oxidation number), a "secondary valence" corresponding to the coordination number (= bonding). This leads to a general definition of coordination compounds:

> Complexes or coordination compounds are molecules or ions ZL_n in which several uncharged or charged, mono- or polyatomic groups = ligands **L** are attached to an uncharged or charged central atom **Z** in accordance with its coordination number **n.**

This definition is very general. A special feature of complexes is that the central particle has more ligands than would be expected according to its charge or position in the periodic table. This is taken into account in the following definition.

> A complex is a chemical compound in which a central atom is bound to a certain number of binding partners and the central particle binds more binding partners than would be expected according to its charge or position in the periodic table.

Even this definition still has weaknesses. For example, it would lead to NH_4^+ or H_3O^+ being a complex by definition because they have more binding partners (H^+) than would be expected on the basis of their charge (N^{-3} or O^{-2}). It becomes even more precise if the properties of the central atom (a Lewis acid) and the ligands (Lewis base) are taken into account (see Sect. 1.2). Refined, the definition then reads:

> A complex is a chemical compound in which a central atom, which is a Lewis acid, is bound to a certain number of binding partners (Lewis bases) and the central atom binds more binding partners than would be expected according to its charge or position in the periodic table.

1.2   Bonding

In order to get a first approach to the bonding relationships in complexes, we first review the basic properties and differences between a covalent bond and ionic bonding.

- Covalent bonds are present in many molecules and some solids. Simple examples are diatomic molecules such as H_2 or Cl_2. Both atoms make one of the valence electrons available for forming a shared electron pair. Since both atoms are fully assigned the shared electron pair, each achieves a noble gas configuration. The bond is directional. $Cl + Cl \rightarrow Cl\text{-}Cl$ (Fig. 1.6 above).
- Ionic bonding is present in salts (e.g. NaCl). Here, too, both atoms reach a noble gas configuration. In contrast to the covalent bond, there is a high electronegativity difference between the two reaction partners (by definition $\Delta E_N > 1.7$). Therefore, one reactant donates one (or more) electron(s) and the other partner accepts it (them). In the case of NaCl, the sodium (electron configuration [Ne] $3\,s^1$) donates an electron and the cation Na^+ with the noble gas configuration [Ne] is formed. The chlorine atom has the electron configuration [Ne] $3\,s^2\,3p^5$; the addition of one electron gives the chloride anion Cl^- with the noble gas configuration of argon. Attractive electrostatic interactions are present between the anion and the cation. These are not directional (Fig. 1.6 centre).
- In coordinative bonding, the ligand serves as an electron pair donor – it is therefore a Lewis base. With this, we have already established the most important property of ligands: They must have at least one free electron pair. The central particle is an electron deficient compound, i.e. a Lewis acid. The coordinative bond can be described as a Lewis acid-base bond. Caution. Not every Lewis acid-base bond is a coordinative bond ($H_3C^+ + |CH_3^- \rightarrow H_3C\text{-}CH_3$). We will see later that even in this bond, the central atom aims to get noble gas configuration. Similar to a covalent bond, there is a common electron pair that is shared by both reactants, resulting in both reaching a noble gas configuration. Alternatively, the coordinative bond can also be explained by attractive electrostatic interactions between a positively charged central atom and negatively charged ligands. However, here the interactions are directional compared to ionic bonding. As an example, we have already considered the reaction of the Lewis acid aluminum (3+) with the Lewis base hydroxide (Fig. 1.6 below). Three covalent coordinative bonds and one coordinative bond are formed.

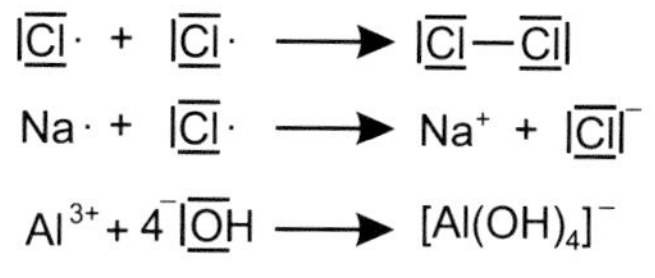

$$|\overline{Cl}\cdot \; + \; |\overline{Cl}\cdot \; \longrightarrow \; |\overline{Cl}-\overline{Cl}|$$
$$Na\cdot \; + \; |\overline{Cl}\cdot \; \longrightarrow \; Na^+ \; + \; |\overline{Cl}|^-$$
$$Al^{3+} + 4\,|\overline{O}H \; \longrightarrow \; [Al(OH)_4]^-$$

Fig. 1.6 Example of the formation of a covalent bond (top), an ionic bond (middle) and a coordinative bond (Lewis acid-base adduct, bottom). In all cases, the reactants involved reach a closed electron shell

In complexes, the ligands coordinate to form a Lewis acid-base bond to the metal center, which is why complexes are also called coordination compounds. We repeat the difference between Lewis and Brønsted acids and bases. According to the Brønsted definition, an acid is a proton donor and a base is a proton acceptor. The oxonium ion H_3O^+ can donate a proton and is therefore an acid, and the hydroxide ion OH^- can accept a proton and is therefore a base. Water (H_2O) can react both as an acid and as a base. Such substances are called ampholytes. In the definition according to Lewis, acids are electron-deficient compounds and bases have free electron pairs. In this case, the acid is the proton H^+, the base is, as in the definition according to Brønsted, the hydroxide ion, which has free electron pairs.

To better highlight the difference between a covalent bond and a covalently coordinative bond, we compare the C-C bond in ethane (H_3C-CH_3) with the N-B bond in ammine borane ($H_3N \rightarrow BH_3$). Both compounds are isoelectronic, meaning that they are analogous in structure and have the same number of atoms and valence electrons; in this particular case, the total electron number is also the same. In both compounds, a directional bond is present. Both the carbon atom and the nitrogen atom are sp^3 hybridized, there are four sp^3 hybrid orbitals present. In the case of the carbon atom with four valence electrons, the four orbitals are singly occupied. Three of them form a covalent bond with the singly occupied s orbitals from the hydrogen atoms, the fourth is available for the covalent bond between the two carbon atoms. In all bonds, pairing of previously unpaired electrons occurs.

The nitrogen atom in amminborane has five valence electrons. Of the four sp^3 hybrid orbitals, three are singly occupied and one is doubly occupied with electrons. The three single-occupied hybrid orbitals are responsible for the three covalent bonds with the single occupied s orbitals of the hydrogen atoms. A free pair of electrons is present in the fourth hybrid orbital. The ammonia molecule is a Lewis base. The boron atom has one valence electron less than the carbon atom. The three valence electrons are located in three sp^2 hybrid orbitals, which are responsible for the three covalent bonds to the three hydrogen atoms. The third p orbital on the boron atom is empty. In the ammonia molecule, the formation of three covalent bonds to the three hydrogen atoms gives the nitrogen atom a total of four electron pairs (three electron pairs in covalent bonds + one free electron pair), giving it a self-contained noble gas shell with eight valence electrons. The boron atom in borane has only the three electron pairs from the three covalent bonds. It has only six valence electrons and is therefore an electron-deficient compound, a Lewis acid. In amminborane, the nitrogen atom provides its free pair of electrons to the boron atom to fill up its valence electron number. A covalent coordinative bond is formed in which the Lewis base ammonia provides the shared electron pair. The boron atom of the Lewis acid BH_3 achieves a completed valence electron shell with eight valence electrons through this bond. The differences additionally exist in the bond cleavage. The C-C bond is homolytically cleaved, and the shared electron pair becomes two unpaired electrons again. The B-N bond is cleaved heterolytically, the common electron pair of the B-N bond remains with the nitrogen atom. This does not necessarily mean that covalent bonds are always homolytically cleaved.

To distinguish a coordinative bond from a covalent bond, the latter is often marked with an arrow (as shown in the case of amminborane) or shown as a dashed line. In many books, no distinction is made and both bonds are shown as a solid line. This variant is also used in this textbook.

1.3   Questions

1. Define the term complex as generally as possible! Use already known concepts here!
2. Name typical properties of ligands!
3. Consider which of the following compounds can be regarded as complexes! Apply the different definitions for a complex! H_2O, H_3O^+, NH_4^+, NH_2^-, ICl, ICl_3, SF_6, SO_4^{2-}, MnO_4^-, $BeCl_4^{2-}$, $BeCl_2(OR_2)_2$ (R = organic subsituent).
4. Look for structures of biologically important transition metal compounds (complexes) that you can approximate now!
5. State the properties of a Lewis base and a Lewis acid. What is meant by a Lewis acid-base adduct?
6. Compare the coordinative bond with the ionic bond and the covalent bond. What are the similarities and differences?
7. Why is the most stable oxidation state of cobalt in aqueous medium +2?

From the beginning of coordination chemistry, the color of coordination compounds had fascinated researchers and the variety of colors has been reflected in the naming of the new compounds. With the multitude of complexes known today, this criterion for nomenclature is no longer sufficient and new rules are necessary to clearly describe the different complexes and their structures.

2.1   IUPAC Nomenclature of Coordination Compounds

The IUPAC (International Union of Pure and Applied Chemistry) regularly issues guidelines for the nomenclature of compounds. In 2005, new recommendations for the nomenclature of inorganic compounds were issued, which also concern coordination chemistry. In the case of complexes, there are rules for the establishment of complex formulae and rules for the systematic naming of complexes [12].

2.1.1   Establishment of the Formulae of Coordination Compounds

A complex consists of the central atom and a certain number of ligands. The examples from the introduction have already shown that complexes can be anionic, cationic or neutral. These four items must be included in the complex formula: central atom, ligands, number and charge. To distinguish complexes from other compounds, the coordination entity is written in square brackets, and the charge, if present, is given as an exponent. In the brackets, the central atom comes before the ligands. The latter follow in alphabetical order, with abbreviations (e.g. py for pyridine or en for ethylenediamine) treated in the same way

© The Author(s), under exclusive license to Springer-Verlag GmbH, DE, part of
Springer Nature 2023
B. Weber, *Coordination Chemistry*,
https://doi.org/10.1007/978-3-662-66441-4_2

as formulas. For ease of reference, abbreviations and polyatomic ligands are given in round brackets. We look again at the examples from the introduction:

$[Al(OH)_4]^-$; $[Ag(NH_3)_2]^+$; $[Cu(H_2O)_6]^{2+}$; $[CuCl_2(H_2O)_4]$; $[Cu(H_2O)_2(NH_3)_4]^{2+}$

The rules were followed in all cases. Under certain conditions, it may be possible to have two or more isomers for the same coordination entity. An example from the introduction is the cobalt(III) chloride complex with four ammonia ligands or two ethylenediamine (en) ligands, respectively, where a green (praseo) and a violet (violeo) variant have been isolated. The two complexes differ in the position of the two chloride ions bound to the cobalt with respect to each other. In the violet case, they are side by side, that is, they are *cis*-standing, i.e., *cis*-$[CoCl_2(en)_2]Cl$, whereas in the green compound they are opposite each other, that is, they are *trans,* i.e., *trans*-$[CoCl_2(en)_2]Cl$. Such structural descriptors precede the complex formula, they are italicized and are connected to the formulae by a hyphen. The rules are summarized again below:

- Coordination entity in square brackets, if necessary charge as an exponent
- Central atom before ligands
- Alphabetical order for ligands (abbreviations like formulas)
- Polyatomic ligands and abbreviations in round brackets
- Oxidation number as exponent behind the central atom
- Structural descriptors before the complex formula with the help of prefixes *cis-, trans-, fac-* (facial), *mer-* (meridional).

2.1.2   Names of Coordination Compounds

The nomenclature concerns the systematic name of the complexes, i.e. how we pronounce or write out the name. We do *not* write out $[Al(OH)_4]^-$ as in the formula – this name would mislead the listener or reader. The systematic name is tetrahydroxidoaluminate(III) or tetrahydroxidoaluminate(1-). We see that, unlike the formula, the ligands are given before the central atom and their number is prefixed in Greek numerals – this is analogous to organic chemistry, where the number of substituents is given in the same way. An example would be 1,2-dichloroethane.

   In order to avoid confusion in the case of more complex ligand names, these are indicated in brackets and provided with the prefixes bis, tris, tetrakis, ... A good example are the two copper complexes $[Cu(H_2O)_2(NH_3)_4]^{2+}$ and $[Cu(CH_3NH_2)_4(H_2O)_2]^{2+}$. In the first complex, water (aqua) and ammonia (ammine) are the ligands and the name is tetraamminediaquacopper(II) – both are simple ligands and the name is unique. In the second complex, methylamine was used as the ligand instead of ammonia. Now the name is diaquatetrakis(methylamine)copper(II). If we had written tetramethylamine, we could also refer to the cation $N(CH_3)_4^+$.

From the examples given so far, we see that no spaces are left between the components of a coordination entity. The ligands are consistently given in alphabetical order – though the order may change compared to the coordination compound formula! For anionic ligands the anion ending is changed to -*o* (e.g. hydroxido), for neutral or formally positively charged ligands there is no denoting ending (e.g. methylamine). Neutral ligands are usually enclosed in round brackets. There are exceptions to this rule, namely the frequently used ligands ammine (NH_3), aqua (H_2O), carbonyl (CO) and nitrosyl (NO). For these ligands, "special" names are also used – i.e. aqua instead of water.

When the naming of the ligands is complete, we continue with the name of the central atom. In the case of anionic complexes, there is again a denoting suffix – in this case -*ate*. We have already seen this in the first example, the tetrahydroxidoaluminate(III) or tetrahydroxidoaluminate(1-). With this, we are almost there. The charge of the complex is still missing, which comes in round brackets after the name. Alternatively, the oxidation state of the metal atom can be given in round brackets with Roman numerals behind the name – the charge of the complex is given with Arabic numerals. What is given does not matter – from the oxidation state of the metal center the charge of the complex can be determined and vice versa. In this book, apart from the introductory examples, the oxidation state of the metal atom is consistently given.

All that is missing now are the additional structural descriptors, which are given before the name of the complex, just as for the formula. Our two cobalt complexes are called *cis*- and *trans*-tetraamminedichloridocobalt(III) chloride, respectively. In the following, the rules are briefly summarized again and Table 2.1 gives an overview of the names of simple, frequently used anionic and neutral ligands. Table 2.2 summarizes the naming of metal atoms in anionic complexes. It should be briefly noted that prior to the 2005 revision of the rules for the IUPAC nomenclature of inorganic compounds, some ligands were named differently, and these names can still be found in many textbooks as well as in the

Table 2.1 Overview of the names of frequently used anionic and neutral ligands. The old formulation corresponds to the rules before 2005, which can still be found in many textbooks, but should no longer be used

Anionic ligands			Neutral ligands	
Abbreviation	Old	New	Abbreviation	
F^-	Fluoro	Fluorido	H_2O	Aqua
Cl^-	Chloro	Chlorido	NH_3	Ammine
OH^-	Hydroxo	Hydroxido	CO	Carbonyl
CN^-	Cyano	Cyanido	NO	Nitrosyl
NCO^-		Cyanato		
O^{2-}	Oxo, oxido	Oxido		
NH_2^-		Amido		
H^-		Hydrido		

Table 2.2 Naming of metal atoms in anionic complexes

Sc	-scandate	La	-lanthanate	Ac	-actinate
Ti	-titanate	Zr	-zirconate	Hf	-hafnate
V	-vanadate	Nb	-niobate	Ta	-tantalate
Cr	-chromate	Mo	-molybdate	W	-tungstate
Mn	-manganate	Tc	-technetate	Re	-rhenate
Fe	-ferrate	Ru	-ruthenate	Os	-osmate
Co	-cobaltate	Rh	-rhodate	Ir	-iridate
Ni	-nickelate	Pd	-palladate	Pt	-platinate
Cu	-cuprate	Ag	-argentate	Au	-aurate
Zn	-zincate	Cd	-cadmate	Hg	-mercurate

Table 2.3 Examples illustrating the nomenclature of complexes

Complex formula	Systematic name
$Na[Al(OH)_4]$	Sodium tetrahydroxidoaluminate(III) or sodium tetrahydroxidoaluminate(1-)
$K[CrF_5O]$	Potassium pentafluoridooxidochromate(VI) or potassium pentafluorido-oxidochromate(1-)
$(NH_4)_2[PbCl_6]$	Ammonium hexachloridoplumbate(IV) or ammonium hexachloridoplumbate (2-)
$[CoCl(NH_3)_5]Cl_2$	Pentaamminechloridocobalt(III) chloride or pentaamminechloridocobalt(2+) chloride
$[Al(H_2O)_5(OH)]Cl_2$	Pentaaquahydroxidoaluminium(III) chloride or pentaaquahydroxidoaluminium (2+) chloride
$[PtCl_4(NH_3)_2]$	Diamminetetrachloridoplatinum(IV) or diamminetetrachloridoplatinum(0)

older literature. In Table 2.3, the nomenclature rules are again clarified by some further examples.

- Ligands in alphabetical order before the name of the central atom, regardless of the number. The number of ligands is prefixed in Greek numeral words.
- The prefixes di, tri, tetra, penta, hexa ...are used for simple ligands, the prefixes bis, tris, tetrakis, etc. are used for more complex ligand names, which are then written in brackets. For example, diammine for $(NH_3)_2$, but bis(methylamine) for $(NH_2CH_3)_2$, to distinguish it from dimethylamine $(NH(CH_3)_2)$.
- No spaces are left between parts that belong to the same coordination entity.
- Anionic ligands end in -o, neutral and formally cationic ligands are without a specific ending.
- Neutral ligands are enclosed in round brackets, exceptions: ammine, aqua, carbonyl, nitrosyl.
- After the ligand names follows the designation of the central atom. For anionic complexes the suffix -ate is added.

- Oxidation number of the central atom (Roman numerals) or charge of the coordination unit (Arabic numerals + charge) after the names in round brackets.
- Structural descriptors before the complex names connected with a hyphen using prefixes *cis, trans, fac* (facial), *mer* (meridional). If a ligand can in principle provide different atoms for coordinative bonding with the central atom, the atoms bonded to the central atom are indicated by their italicized symbols after the ligand name, e.g. dithiooxalate dianion ($C_2S_2O_2^{2-}$) is called dithiooxalato-*S,S* if the central atom is coordinated via the two S atoms.

2.2 Nomenclature of Organometallic Compounds

Based on properties and reactivities, there is no clear demarcation between coordination compounds and organometallic compounds. In the following chapter we will learn that organometallic compounds – by definition according to IUPAC – have at least one metal-carbon bond. This definition does not always reflect in the properties of the compound. Classic examples are hexacyanidoferrate(II) and hexacyanidoferrate(III), the complex anions from the so-called yellow and red prussiate of Potash, $[Fe(CN)_6]^{4-}$ and $[Fe(CN)_6]^{3-}$. In both compounds, the cyanide anion coordinates to the iron center via the carbon – thus they possess a metal-carbon bond. However, the properties and bonding ratios are better explained using the concepts of "Werner" complexes. The nickel carbonyl complex from the Mond process for the purification of nickel presented in the introduction, $[Ni(CO)_4]$, the tetracarbonylnickel(0), is a "classical" organometallic compound. This example shows that the same nomenclature rules can be applied to organometallic compounds as for coordination compounds, and IUPAC recommends that we do the same. Consistently, we write the formula in square brackets in this book. However, it is quite common for organometallic compounds to omit these brackets, i.e. to write $Ni(CO)_4$. Please note, this is then not in accordance with the IUPAC rules. The nomenclature used for complexes and also recommended for organometallic compounds according to IUPAC is called additive nomenclature. Alternatively, organometallic compounds can be considered derivatives of organic compounds because of the metal-carbon bond, and nomenclature in accordance with the rules of organic chemistry is possible. This substitutive nomenclature will be demonstrated in the following using only a few selected examples for the sake of completeness.

According to the **additive nomenclature,** organometallic compounds are regarded as coordination compounds stabilized by alkyl or aryl ligands. This nomenclature is mainly used for transition metals, which are also the subject of this textbook. We call the compound $[Ti(C_2H_5)_2(CH_3)_2]$ diethyldimethyltitanium. The complex formula is written in square brackets and the order of the ligand names is alphabetical. Cations and anions are classified by the charge or oxidation number of the central atom in analogy to complexes. Anions are given the suffix *-ate.* $[Ti(C_2H_5)_2(CH_3)]^+$ is diethylmethyltitanium(1+) or –

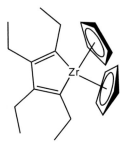

Fig. 2.1 With the substitutive nomenclature, the name is bis(cyclopentadienyl)-2,3,4,5-tetraethyl-1-zircona-cyclopenta-2,4-diene, with the additive nomenclature, bis(cyclopentadienyl)-1:4-η-1,2,3,4-tetraethylbuta-1,3-dienylzirconium(IV)

(IV) and $[Fe(CO)_4]^{2-}$ is tetracarbonylferrate(2-) or -(-II). For organometallic compounds, it is often difficult to accurately determine the oxidation state of the metal center. In these cases, it is advisable to give the total charge of the organometallic entity. It is always unambiguous.

In **substitutive nomenclature,** organometallic compounds are regarded as organometallics, i.e. as derivatives of binary hydrides. In analogy to the alkanes, they are designated with the suffix *-ane*. This variant is mainly used for main group metals. Since we no longer assume a coordination compound here, the formula is written without square brackets. $(C_2H_5)_2(CH_3)_2Sn$ is diethyldimethylstannane, according to the additive nomenclature the name would be diethyldimethyltin. In substitutive nomenclature, it is also possible to think of organometallic compounds as organic molecules in which metals and also other elements have been substituted in place of carbon. In Fig. 2.1 we see such an example, where zirconium is considered to be part of a five-membered ring.

In the sequel, we consistently apply the additive nomenclature recommended by IUPAC for all organometallic compounds with transition metals. We summarize the most important points once again:

- The same rules are applied to organometallic compounds as to coordination compounds, both in setting up the formulae and in naming them. This is called additive nomenclature.
- Multiple bonds are not taken into account when determining the coordination number.
- If the oxidation state of the metal centre cannot be determined unambiguously, the charge of the entire entity is preferred for the systematic name.

2.2.1   Ligand Names for Organometallic Compounds

The ligands presented so far are anions (OH^-) or neutral compounds (NH_3). For organometallic compounds, the exact assignment of oxidation states for metal and ligand is more difficult and can also be reversed depending on the ligand or metal center used. The decisive factor in determining how the bond is regarded is the difference in electronegativity between the metal center and the coordinated carbon atom. Accordingly, the ligand may coordinate neutrally or as an anion to the metal center. The IUPAC nomenclature distinguishes between the two variants and the following is a brief introduction to each to show the differences. In everyday use, the neutral description has prevailed because of the difficulties in determining the oxidation states precisely, and this will also be used in the further course.

If the organic ligands are considered as **anions,** the ending -ido in analogy to the inorganic anions is used. At first glance, this designation may take some getting used to, but it is very systematic. We compare chlorine and methane: Cl is the chlorine (atom), Cl^- we call chloride and as a ligand in complexes we say chlorido. CH_4 is methane, the corresponding anion CH_3^- is methanide, and the anionic ligand in complexes would then be methanido. Thus the name for the complex $[TiCl_3Me]$ is trichloridomethanidotitanium (IV). Because of the large electronegativity difference between titanium (EN = 1.3) and a sp^3-hybridized carbon (EN = 2.5), this compound can be assumed to have the methyl group as an anion and titanium has the +IV oxidation state.

Another very frequently used anionic ligand for organometallic compounds is the anion of cyclopentadiene, $C_5H_5^-$, the cyclopentadienide. A compound with this ligand, known under the trivial name ferrocene for a very long time, is $[Fe(C_5H_5)_2]$, the bis (cyclopentadienido)iron(II). As mentioned earlier, the assignment of oxidation states is often ambiguous for these compounds. In the literature, the designation of the ligands usually used are for **neutral ligands.** They are then given the identifying suffix -yl. Our two examples are now called trichloridomethyltitanium(IV) and bis(cyclopentadienyl)iron(II). All previous examples have been named with this variant of nomenclature. In principle, both possibilities are correct and IUPAC-compliant, as long as there is nothing to prevent a negative charge on the ligand. In Table 2.4, the systematic names as anionic and neutral

Table 2.4 Systematic names of selected organometallic ligands as anionic and neutral ligands

Ligand	Name anion	Name neutral	Alternative name
CH_3-	Methanido	Methyl	
CH_3-CH_2-	Ethanido	Ethyl	
CH_3-CH_2-CH_2-	Propane-1-ido	Propyl	
$(CH_3)_2$-CH-	Propane-2-ido	Propan-2-yl 1-methylethyl	Isopropyl
CH_2=CH-CH_2-	Prop-2-en-1-ido	Prop-2-en-1-yl	Allyl
C_6H_5-	Benzenido	Phenyl	
C_5H_5-	Cyclopentadienido	Cyclopentadienyl	
CH_3-C(O)-	1-oxoethane-1-ido	Ethanoyl	Acetyl
CH_2=CH-	Ethenido	Ethenyl	Vinyl

Table 2.5 Examples illustrating the nomenclature of organometallic compounds

Complex formula	Systematic name
[OsEt(NH$_3$)$_5$]Cl	Pentaammine(ethyl)osmium(+1) chloride
Li[CuMe$_2$]	Lithium dimethylcuprate(−1)
[Rh(py)(PPh$_3$)$_2$(C≡CPh)]	(Phenylethynyl)(pyridine)bis(triphenylphosphane)rhodium(0)

ligand are given again for a few examples. In Table 2.5, the nomenclature rules are clarified by means of a few further examples.

2.3   Information on the Structure

Especially in organometallic compounds, there are often many different variations of how a ligand can coordinate to the metal. While in coordination compounds the donor atoms of the ligands are often unique, in organometallic compounds all or only selected carbon atoms can coordinate to the metal center and the mode of binding can also change or differ in one compound. To account for this, there are additional designations of varying significance that can and should be applied depending on the complexity of the structure. Such structural indications are important not only for organometallic compounds, but also for many coordination compounds, especially when several metal centers are present in a complex, additional information is necessary to deduce the actual spatial structure from the names. In this book, the μ-convention and the η-convention are introduced. The κ-convention, which is well suited for much more complex systems, is briefly introduced.

2.3.1   The μ Convention

In both coordination compounds and organometallic compounds, ligands frequently occur that have two or more donor atoms. These can all coordinate to the same metal atom. In this case, the ligand acts as a chelating ligand. Alternatively, two or more metal atoms may be bound to one ligand. In this case, a bridging ligand is present. This bridging mode of binding of a ligand is denoted by the Greek letter μ. The number of metal centers bridged by the ligand is indicated by a subscript n ($n \geq 2$), with the subscript 2 usually omitted. The propane-derived ligand –CH$_2$-CH$_2$-CH$_2$– can coordinate to the same metal center as a chelate ligand via the 1- and 3-positions. In that case, the name is propane-1,3-diyl. If two metal centers are bridged, the name is μ-propane-1,3-diyl.

Monatomic ligands can also serve as bridging ligands. Thus, complexes with μ-hydrido or μ-halogenido ligands are not uncommon. Three examples follow to illustrate these rules. The bridging ligand is usually given first.

The first example (Fig. 2.2a) is a dinuclear cobalt carbonyl complex with two bridging carbonyl ligands. The metal-metal bond occurring here is indicated after the name with additional parenthesis and in italics. The name is bis(μ-carbonyl)bis[tricarbonylcobalt(0)]

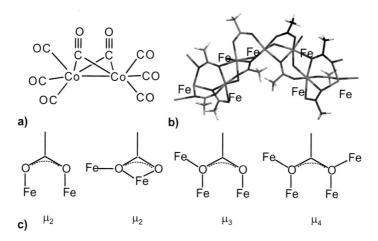

Fig. 2.2 (**a**) Structure of bis(μ-carbonyl)bis[tricarbonylcobalt(0)] *(Co-Co)*. (**b**) Detail of the crystal structure of iron(II) acetate. (**c**) Different bridging modes of the acetato ligand [13]

(Co-Co). Two cobalt atoms, each with three monodentate carbonyl ligands, are held together by two bridging carbonyl ligands and a cobalt-cobalt bond.

The second example in Fig. 2.2 is a section of the solid state structure of iron(II) acetate. In the molecular formula, there are two acetate ions per iron and the coordination number of the iron(II) ion is six. This only works because the acetate ions bridge two to four iron ions, with the oxygen atoms either being monodentate or forming a μ_2-bridge. The subscript locant "counts" the number of bridged metal centers per acetate unit. Figure 2.2c shows the four different bridging modes realized in the solid state. They can be classified as μ_2-acetato, chelating μ_2-acetato, μ_3-acetato and μ_4-acetato.

The third compound is a decanuclear thorium compound named ditriakontaamminehexadeca-μ-fluoridotetra-μ_3-oxidotetra-μ_4-oxidodecathorium(IV) and the complex formula $[Th_{10}(\mu\text{-}F_{16})(\mu_3\text{-}O_4)(\mu_4\text{-}O_4)(NH_3)_{32}]^{8+}$ (Fig. 2.3). The thorium atoms are μ_4-bridged by four oxygen atoms. Thereby, six thorium atoms form an adamantane framework with the four oxygen atoms. The same structural motif is also found in the diamond structure and the phosphorus oxide structures (P_4O_6 and P_4O_{10}). The four remaining thorium atoms occupy the fourth position above each of the oxygen atoms (Fig. 2.3, far left). Three thorium atoms from the adamantane backbone are always μ_3-bridged by a total of four oxygen atoms and the outer thorium atoms are connected to the backbone by the μ_2-bridging fluoride ions (Fig. 2.3 middle). The coordination sphere around each thorium is now saturated by a total of 32 ammonia molecules. In the process, the thorium reaches coordination numbers 9 and 10, which are rare but not unexpected for such a large ion (Fig. 2.3 right).

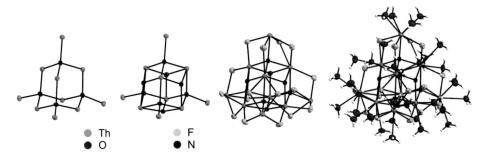

Fig. 2.3 Systematic structure of a decanuclear thorium compound with μ_4- and μ_3-bridging oxide ions and μ-bridging fluoride ions [14]

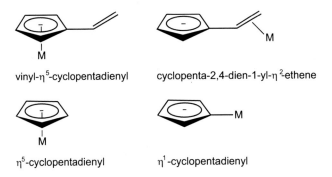

Fig. 2.4 Possible binding modes of unsaturated hydrocarbons using the example of 1-ethenylcyclopenta-2,4-dien-1-yl (top) and cyclopenta-2,4-dien-1-yl (bottom)

2.3.2 The η Convention

Unsaturated hydrocarbons as ligands can coordinate to the central atom via the π-electrons instead of the free electron pair discussed so far. Different modes are possible, which can be distinguished using the "hapto" nomenclature. As an example, we consider 1-ethenylcyclopenta-2,4-dien-1-yl. The metal can be coordinated via the π-electrons on the ethene or on the cyclopenta-2,4-dien-1-yl. In the first variant, the ligand would be the cyclopenta-2,4-dien-1-yl-η^2-ethene and in the other, the vinyl-η^5-cyclopentadienyl. Both variants are given in Fig. 2.4.

The number of unsaturated C atoms coordinated to the metal is indicated by the superscript number after the Greek letter η (eta). η^n is then read as n-hapto, i.e. trihapto (n = 3), tetrahapto (n = 4), etc. If all C atoms coordinate in benzene, then it is bound as η^6-benzene. Similarly, a cyclopentadienyl coordinated over all five C atoms is bonded as η^5-cyclopentadienyl. If a σ bond is formed to one of the C atoms instead of the π bonds, we refer to the ligand as σ-cyclopentadienyl or η^1-cyclopentadienyl. If not all unsaturated centers of a ligand are involved in a bond or if a ligand allows several bond variants, the

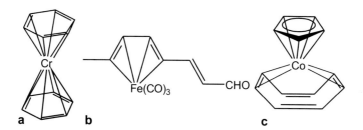

Fig. 2.5 (**a**) Di(η^6-benzene)chromium(0), (**b**) Tricarbonyl(4–7-η-octa-2,4,6-trienal)iron(0) and (**c**) (1,2:5,6-η-cyclooctatetraene)-(η^5-cyclopentadienyl)cobalt(I)

corresponding locants, i.e. numbers describing the atom, are introduced before the character η. A hyphen can be used to group together several consecutive C atoms. If some are omitted in between, this is made clear with a colon between the locants, as can be seen in the following examples in Fig. 2.5.

2.3.3   The κ Convention

With increasing complexity of the molecules, the κ convention has proven itself: m κ^n *element symbol*^{l} is placed after the ligand name (with m = locant, which "counts" the metal centres in multinuclear complexes; n = locant, which indicates the number of identical ligator atoms; l = locant, which contains the numbering of the coordinating atom and the element symbol of the ligator atom written in italics). Here, l is of particular importance. In the compound shown in Fig. 2.6, there is only one metal center, so the locant m is omitted. There are no identical ligator atoms per ligand, so n is also omitted. In this case, the κ convention is limited to indicating that one of the two triphenylphosphane ligands coordinates only via the P atom on the nickel, while for the second triphenylphosphane the ligand additionally coordinates via C^1 from one of the three phenyl rings.

The information on the structure can be given in the name as well as in the formula of the compound, as shown in Fig. 2.7 for the example of aluminium trichloride occurring as a dimer.

Fig. 2.6
[2-(diphenylphosphanyl-κP)-
phenyl-κC^1]hydrido
(triphenylphosphan-κP)nickel
(II)

Fig. 2.7 [Al$_2$Cl$_4$(μ-Cl)$_2$] or
[Cl$_2$Al(μ-Cl)$_2$AlCl$_2$] or Di-μ-
chlorido-tetrachlorido-
1κ^2Cl,2κ^2Cl-dialuminium

2.4   Structure of Complexes

The coordination compounds are so numerous and so different in their composition, their properties and their chemical behaviour that it would be quite impossible and also completely pointless to try to memorise all the complexes. It is much more important to find *structural features* and properties that allow a *classification of* the complexes and to bring an order, a system, into this seemingly so confusing area. One way to systematize complexes is to group them according to the coordination number and the corresponding coordination polyhedra. Another way is to sort the ligands. In this case the concept of denticity has proven to be very useful. In the following, both concepts are presented, since the properties of complexes are strongly influenced by the coordination number and the type of ligands, in addition to the central atom.

2.4.1   Ligands and Their Denticity

Ligands are molecules or atoms with at least one free electron pair. This definition applies to a very broad spectrum of compounds. It therefore makes sense to introduce an additional subdivision of ligands. Here, the concept of denticity has proved particularly useful. This describes the number of donor atoms in a ligand that coordinate to a metal center. Of the different ligand types thus obtained, multidentate ligands can be further divided into categories such as *linear* (e.g. diethylenetriamine, diene), *branched* (e.g. edta) and *macrocyclic* (e.g. phthalocyanine). Some of the monodentate ligands we have encountered during our studies have already been introduced in the introduction (e.g. hydroxide, cyanide, carbon monoxide). Of the multidentate ligands, the anion of ethylenediaminetetraacetic acid (edta) is perhaps already known from quantitative analysis (complexometric titration, complexometry) as well as because of its importance in heavy metal detoxification.

Monodentate Ligands are ions or molecules which, although sometimes possessing several free electron pairs, only form one bond to one metal centre. Examples are halide and pseudohalide anions, water, ammonia, aliphatic alcohols, amines. If several free electron pairs are present, monodentate ligands can form multiple bonds or act as bridging ligands for two metal centers. This is then marked with the already introduced μ convention.

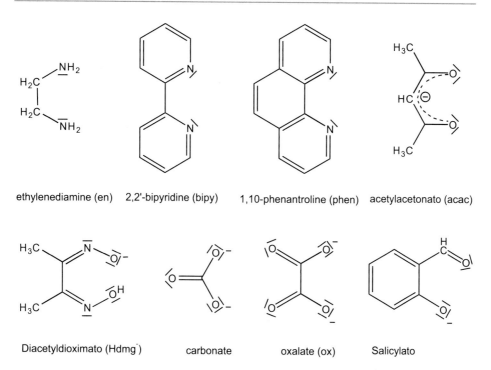

ethylenediamine (en)   2,2'-bipyridine (bipy)   1,10-phenantroline (phen)   acetylacetonato (acac)

Diacetyldioximato (Hdmg⁻)   carbonate   oxalate (ox)   Salicylato

Fig. 2.8 Selected bidentate ligands

Bidentate Ligands are ions or molecules that have multiple free electron pairs, two of which are used for the coordination to a metal center. The examples shown in Fig. 2.8 are bidentate ligands, which may have occurred in the course of chemistry studies. This would be, for example, the diacetyldioximato anion that is used in the quantitative determination of nickel. The bis(diacetyldioximato)nickel(II) complex is characterized by a pronounced stability. This is achieved, among other things, by additional stabilizing hydrogen bonds between the two coordinating ligands. Figure 2.9 shows a schematic diagram of the complex. A prerequisite for the high stability is the square planar structure of the complex, which is not observed for the corresponding cobalt(II) complex, for example. The high stability and low solubility of the bis(diacetyldioximato)nickel(II) complex allows quantitative detection of nickel alongside cobalt, something that is not possible with edta, a six-dentate ligand. Why the nickel(II) complex is square planar in this particular case is explained in the chapter binding models (ligand field theory).

The bidentate ligands shown act as chelating ligands, thus both bonds are formed with the same metal center. The term chelate ligand comes from the Latin chelae or the Greek chele, both meaning crab claws.

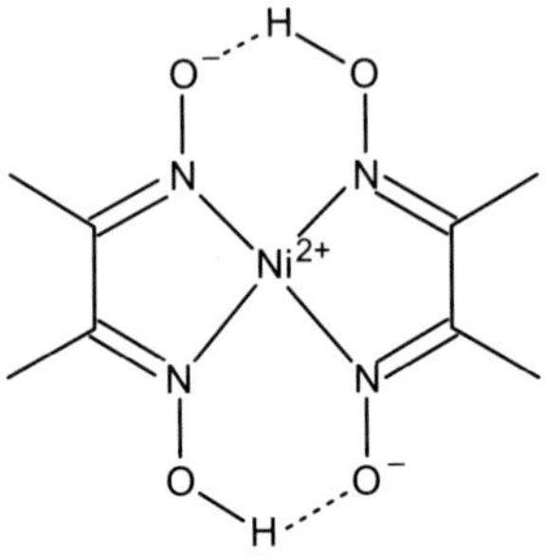

Fig. 2.9 Structure of the bis
(diacetyldioximato)nickel
(II) complex. The intramolecular
hydrogen bonds lead to a
significant increase in the
stability of the square planar
complex, which is used as a very
selective nickel detector

Complexes with multidentate ligands which form closed, n-membered units = chelate
rings with the central atom (preferably 5- and 6-membered, 5-membered are particularly stable) are called chelate complexes.

Tridentate Ligands
provide three free electron pairs for coordination to a metal center. Commonly used
examples are the trispyrazolyl borate and the terpyridine shown in Fig. 2.10. The two
ligands, because of their rigide structure, can force different isomeric forms in octahedral
complexes: a *fac-* (trispyrazolyl borate) or *mer-* (terpyridine) conformation (see Isomerism).

Tetradentate Ligands Among the tetradentate ligands, macrocyclic ligands play an
important role in addition to linear ligands such as polyamines or pyridine derivatives. A
selection is given in Fig. 2.11. Most were prepared to model biologically relevant
macrocycles such as protoporphyrin IX from hemoglobin (red blood pigment and relevant

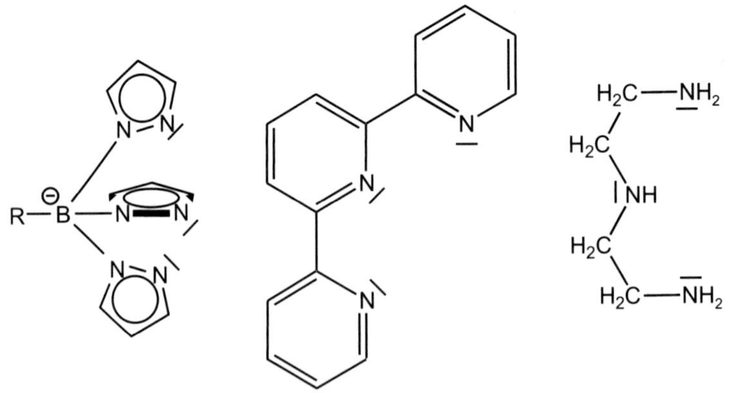

Trispyrazolylborate (tp) 2,2':6',2''-Terpyridine (terpy) Diethylenetriamine (dien)

Fig. 2.10 Selected tridentate ligands

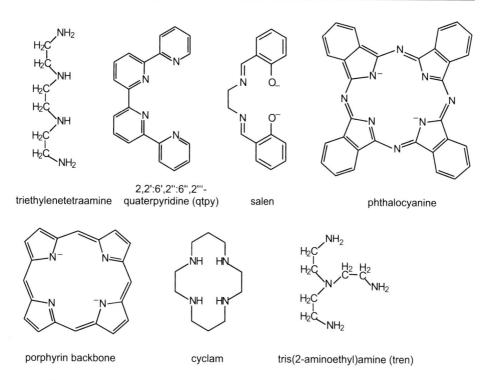

Fig. 2.11 Selected tetradentate linear, branched (tren) and macrocyclic ligands (porphyrin, phthalocyanine, cyclam)

for oxygen transport in our body). Many biologically relevant catalytic processes are based on complexes with macrocyclic ligands. In part, the modelling of such processes is also possible with linear ligands, including Schiff bases such as the Salen ligand.

Pentadentate Ligands are rare compared to the examples discussed so far. Many complexes with coordination number five have one tetradentate and one monodentate ligand. An example is given in Fig. 2.12.

A classic example of a **hexadentate ligand** is the ethylenediaminetetraacetate (edta) anion, a non-selective ligand that forms stable 1:1 complexes with a variety of one- to four-times charged metal ions. The complex formation is strongly pH-dependent. The tetra-anion of ethylenediaminetetraacetic acid acts as the ligand and the concentration of this anion in solution is strongly dependent on its pH.

Many ligands used in coordination chemistry are too complicated to be included in a complex formula in an unambiguous way with their molecular formula. For these ligands, abbreviations are usually used similar to those used in organic chemistry for certain

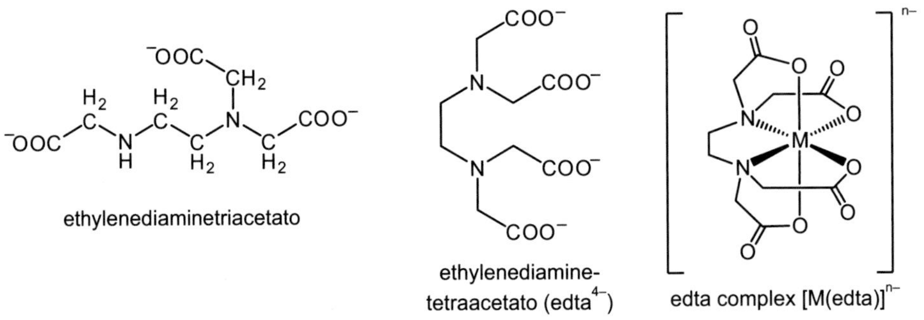

ethylenediaminetriacetato

ethylenediamine-
tetraacetato (edta^{4-})

edta complex [M(edta)]$^{n-}$

Fig. 2.12 Example of a pentadentate ligand (anion of ethylenediaminetriacetic acid) and a hexadentate ligand (anion of ethylenediaminetetraacetic acid, edta^{4-}) and general structure of an edta complex

Table 2.6 Abbreviations of selected ligands

Abbreviation	Ligand name	Abbreviation	Ligand name
acac$^-$	Acetylacetonate	py	Pyridine
thf	Tetrahydrofurane	Hpz	1H-pyrazole
bipy (bpy)	2,2$'$-bipyridine	terpy	2,2$'$:6,2$''$-terpyridine
edta^{4-}	Ethylenediaminetetraacetate	en	Ethylenediamine
tren	Tris(2-aminoethyl)amine	tmeda	Tetramethylethylenediamine
phen	Phenanthroline	dabco	1,4-diazabicyclo[2.2.2]octane
salen^{2-}	Bis (salicylidene)ethylenediamine	cyclam	1,4,8,12-tetraazacyclotetradecane
dmg^{2-}	Diacetyldioximate		

structural groups. Lower case letters should be used for the abbreviations and they should be enclosed in round brackets in the complex formulae. A selection is given in Table 2.6.

2.4.2 Coordination Numbers and Coordination Polyhedra

One of the most important structural features of a complex is the coordination number (CN), i.e. the number of ligands bound to the central atom or, in the case of multidentate ligands, the number of coordinating donor atoms. The actual bonding situation between metal and ligand does not play a role in the determination of the coordination number. Especially in organometallic compounds, multiple bonds between metal and ligand can potentially occur. For the determination of the coordination number, these potential π and δ bonds are not taken into account. [Ir(CO)Cl(PPh$_3$)$_2$], [RhI$_2$(Me)(PPh$_3$)$_2$] and [W(CO)$_6$] are examples of organometallic compounds with coordination numbers 4, 5 and 6. Closely related to the coordination number is the coordination polyhedron, i.e. the

$$L \text{————} M \text{————} L$$

Fig. 2.13 Coordination number 2

Fig. 2.14 Coordination
number 3

geometrical structure in which the ligands arrange themselves around the central ion. In the
following, the coordination numbers 2–6 and the corresponding coordination polyhedra are
discussed.

Coordination Number 2 Coordination number 2 (Fig. 2.13) is quite rare and restricted to
central ions such as Cu^+, Ag^+ or Au^+. The coordination polyhedron is called *linear*. An
example would be $[Ag(NH_3)_2]^+$.

Coordination Number 3 Coordination number 3 (Fig. 2.14) occurs rarely, and the
corresponding coordination polyhedron is the *trigonal plane*. Examples are $[HgI_3]^-$ and
$[Pt(P(C_6H_5)_3)_3]$.

Coordination Number 4 Coordination number 4 (Fig. 2.15, left) occurs very frequently
in the form of *square planar* and *tetrahedral* complexes. Examples of tetrahedral
complexes are $[Zn(OH)_4]^{2-}$ or $[Cd(CN)_4]^{2-}$; they are often observed for d^0 or d^{10} metal
centers. Examples of square planar complexes are $[PtCl_4]^{2-}$ and $[AuF_4]^-$. This coordination geometry is often observed for d^8 metal centers. Some compounds that would be
expected to have CN 3 by composition occur with CN 4 (Fig. 2.15, right). Two examples

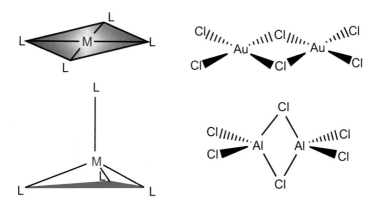

Fig. 2.15 Left: For coordination number 4, the relevant coordination polyhedra are the *square plane*
(top) and the *tetrahedron* (bottom). Right: For some compounds of the composition ML_3 the CN
4 can be achieved by dimerization

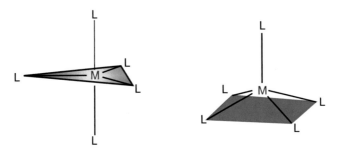

Fig. 2.16 With coordination number 5, the coordination polyhedra are found to be *trigonal bipyramidal* (left) and *square pyramidal* (right)

are gaseous $[(AlCl_3)_2]$ and $[(AuCl_3)_2]$, where CN 4 is realized by dimerization. In the case of aluminium the coordination environment is tetrahedral and in the case of gold it is square planar, which again can be explained by the *d*-electron number.

Coordination Number 5 Coordination number 5 is relatively rare and occurs in the form of *trigonal bipyramidal* and *square pyramidal* complexes (Fig. 2.16). The positions occupied by the ligands above and below the plane in the trigonal bipyramid are called axial, and the positions occupied by the three ligands in the plane are called equatorial. Examples of trigonal bipyramidal complexes are $[SnCl_5]^-$ and $[Fe(CO)_5]$; an example of a square pyramidal complex is $[VO(acac)_2]$ with acac $=$ acetylacetonate. The energetic differences between the trigonal bipyramidal and the square pyramidal are small and both can be transformed into each other by minor deformation. In trigonal bipyramidal complexes, the axial and equatorial positions are not equivalent and can be transformed into each other by deformation vibrations. As an intermediate step, a square pyramidal arrangement is obtained. This process is called Berry pseudorotation mechanism.

Coordination Number 6 With the coordination number 6 the coordination polyhedra octahedron, trigonal prism, trigonal antiprism or planar hexagon are conceivable (Fig. 2.17 a). In numerous experiments *A. Werner* and his co-workers could prove that only the octahedron occurs, unless steric requirements of the ligand dictate another structure. The octahedron is a special case of the trigonal antiprism, where all edges are of equal length (Fig. 2.17b). However, the ideal octahedron is rare and deviations from the ideal octahedral symmetry (often stretched or compressed along the four-fold axis) occur in most octahedral complexes. Increasing distance of the ligands along the *z*-axis leads to the lowering of the CN from 6 to 4 and by this to a square planar geometry. Some compounds that would be expected to have CN 5 by composition occur with CN 6. Two examples for such a behavior are the dimeric $[Mn_2I_2(CO)_8]$ and the tetrameric $[(MoF_5)_4]$ (Fig. 2.17c).

Higher Coordination Numbers With higher coordination numbers, several possible coordination polyhedra always occur, some of which differ only slightly from each

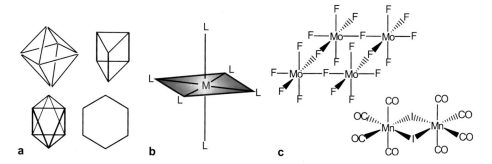

Fig. 2.17 (**a**) With coordination number 6, the coordination polyhedra octahedron, trigonal prism, trigonal antiprism or planar hexagon are conceivable. (**b**) All complexes with coordination number 6 occur as octahedra. (**c**) For some compounds of the composition ML_5 the CN 6 can be realized by the formation of larger aggregates

other. In general, higher coordination numbers are rare and are only found with large central ions. Examples of this are complexes of lanthanides and actinides, such as the thorium cluster mentioned earlier (Fig. 2.3), where coordination numbers 9 and 10 occur.

2.5   Isomerism in Coordination Compounds

Isomerism is the phenomenon that chemical compounds exist with the same composition but different arrangement of atoms. This phenomenon also occurs in coordination chemistry in various forms, some of which are briefly discussed here. The first examples are constitutional (structural) isomers with different linkage sequences of the atoms. Typical examples of this from organic chemistry would be *n*-octane/isooctane or *ortho−/meta/para*-nitrotoluene. The isomers have different chemical and physical properties.

Ionization Isomerism Many anions can be a ligand in the complex or a counter ion outside the complex, so that different ions with often very different properties are present in the solution.

$[CoCl(NH_3)_5]SO_4$ and $[Co(NH_3)_5SO_4]Cl$

Hydrate or Solvate Isomerism Hydrate isomerism is a special form of ionization isomerism, where H_2O or other solvent molecules are exchanged for other ligands. A well-known example is $CrCl_3 \cdot 6\ H_2O$, of which four different isomers are known. An aqueous solution of $CrCl_3 \cdot 6\ H_2O$ is initially dark green, but on standing it slowly becomes lighter, changing colour through a pale blue-green to violet. Here the different isomers occur in the

Table 2.7 Ligand names of selected linkage isomers

Structure	Ligand name
M—$\bar{\text{S}}$—C≡N|	thiocyanato-
$\bar{\text{I}}\bar{\text{S}}$—C≡N—M	isothiocyanato-
M—$\bar{\text{C}}$≡N|	cyanido-
$\bar{\text{I}}$C≡N—M	isocyanido-
M—$\bar{\text{O}}$—$\bar{\text{N}}$=$\bar{\text{O}}$	nitrito-O-
M—N with O^- and O	nitrito-N-

following order, the first two being green, the third blue-green, and the fourth violet. The process can be reversed again by heating the solution.

$$[CrCl_3(H_2O)_3]\cdot 3\ H_2O \rightleftharpoons [CrCl_2(H_2O)_4]Cl\cdot 2\ H_2O \rightleftharpoons [CrCl(H_2O)_5]Cl_2\cdot H_2O \rightleftharpoons [Cr(H_2O)_6]Cl_3$$

Coordination Isomerism If cation and anion are complex ions, different isomers can be obtained by (partially) exchanging the central ion or the ligands. An example would be the salt $[Co(NH_3)_6][Fe(CN)_6]$. A (partial) exchange of the ligands would significantly change the properties of the compound.

Linkage Isomerism The bond of a ligand to the central atom starts from one of the free electron pairs of an atom or molecule. If a ligand has two or more atoms with free electron pairs, the bond to the central ion can originate from different atoms. This ability of a ligand to coordinate to a central ion through different atoms leads to linkage isomerism. The corresponding ligands are called ambidentate (dentate on both sides). Table 2.7 shows the two possible isomers for the ligands thiocyanate, nitrite and cyanide. As a rule, the donor atoms are given in italics after the ligand name. For thiocyanate and cyanide, different names are used for the ligand.

2.5.1 Stereoisomerism

In stereoisomers, the linkage sequence of the atoms is the same and the difference lies in the spatial arrangement. A typical example from organic chemistry is fumaric acid and maleic acid, the *trans-* and *cis-*isomer of butenedioic acid (or ethylenedicarboxylic acid). In complexes, stereoisomers differ in the spatial arrangement of their ligands. The isomers have different chemical and physical properties.

cis-trans **Isomerism** Isomerism occurs in square planar complexes of the general formula $[MA_2B_2]$ and octahedral complexes of the general formula $[MA_2B_4]$. In the *cis-*isomer,

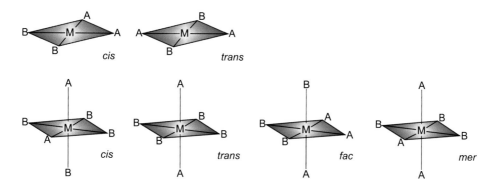

Fig. 2.18 *cis-trans* isomerism for square planar and octahedral complexes and *fac-mer* isomerism for octahedral complexes

identical ligands are adjacent to each other and in the *trans*-isomer they are opposite to each other. In Fig. 2.18, both examples of an octahedral and a square planar complex are shown schematically.

fac-mer **Isomerism** is observed for octahedral complexes with the composition [MA$_3$B$_3$] or octahedral complexes with two tridentate ligands. The prefix *fac* stands for facial (face), that is, identical ligands form a triangular face of the octahedron. With the prefix *mer* for meridional, the same three ligands are each arranged along two adjacent edges of the octahedron. With two tridentate ligands, control over whether facial or meridional coordination occurs can be achieved by choice of ligand. Two examples of this are given in Fig. 2.10. In Fig. 2.18, both variants are shown schematically.

2.5.2   Enantiomers

In organic chemistry, enantiomers are compounds with an asymmetrically substituted carbon atom. A sp^3 hybridized carbon atom with four different substituents can occur in two different isomers, which behave like image and mirror image to each other, but are not convertible into each other. These enantiomers do not differ in their chemical/physical properties except in the sign of the angle of rotation of linearly polarized light. In compounds with two optically active sites (e.g. complex and counterion or complex and ligand), enantiomers or diastereomers are obtained. Diastereomers do not behave like image and mirror image to each other and then differ again in their chemical and physical properties. This circumstance can be used to separate enantiomers.

Optical Isomerism occurs in octahedral complexes with two bidentate and two monodentate ligands or three bidentate ligands and in tetrahedral complexes with four

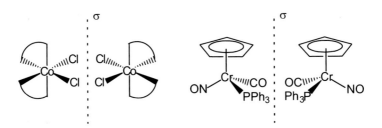

Fig. 2.19 Optical isomerism for the octahedral complex *cis*-[CoCl$_2$(en)$_2$]$^+$ (en = ethylenediamine) and for the tetrahedral complex [Cr(CO)(cp)(NO)(PPh$_3$)] (cp = cyclopentadienyl)

different ligands (analogy to carbon). For octahedral complexes, the Λ-Δ convention (skew-lines convention) was introduced to distinguish the two variants. In the left part of Fig. 2.19, the Λ-isomer (left-handed helical) is shown on the left and the Δ-isomer (right-handed helical) on the right.

2.6   Questions

1. Determine the IUPAC name of the following compounds and ions: [Fe(CO)$_5$], [Ni (CO)$_4$], [Co(H$_2$O)$_6$]$^{2+}$, [Cu(NH$_3$)$_4$(H$_2$O)$_2$]$^{2+}$, [Fe(CN)]$_6^{3-}$.
2. Write the formula of the following compounds and ions: Hexacyanidoferrate(II), Tetraamminplatin(II), Tetrahydroxidozincate(II), Hexacarbonylchromium(0).
3. Name the following complexes according to the IUPAC nomenclature system: [Co (CO$_3$)(NH$_3$)$_4$]NO$_3$, K[MnO$_4$], [CrCl(H$_2$O)$_5$]Cl$_2$, [CrCl$_2$(H$_2$O)$_4$]Cl, Ba[FeO$_4$], Na$_2$[TiO$_3$], [Pt(NH$_3$)$_6$]Cl$_4$, K$_2$[PtCl]$_6$.
4. Give the formulae of the following compounds: Sodium hexafluoridoaluminate(III), Pentacarbonyl iron(0), Potassium hexacyanidoferrate(II), Potassium hexacyanidoferrate (III).
5. Mn(CO)$_4$(SCF$_3$) is dimeric and the manganese is octahedrally surrounded by six ligands in this complex. What structure would you expect? What structures would be expected in a monomeric complex?
6. From the aqueous solution of a complex of the composition CoClSO$_4$· 5NH$_3$ a precipitate is formed with BaCl$_2$, but not with AgNO$_3$. Which complex is present? Are there other isomers and what is the name of this isomeric species?
7. Consider which isomers are to be expected in an octahedral complex [ZA$_3$X$_3$].
8. How do optical isomeric complexes differ? How can they be separated?

What Are Organometallic Compounds?

3

The difference between an organometallic compound and a coordination compound is defined by IUPAC [12, 15]. It states:

> *Organometallic* compounds are characterized by at least one direct bond between a metal atom and a carbon atom.

This rule is clear but can also be confusing. Of the compounds mentioned in the introduction, the metal carbonyls such as the tetracarbonyl nickel(0) of the Mond process are classical organometallic compounds. However, a strict interpretation of the IUPAC definition also leads to classic complexes such as the yellow and red prussiate of Potash [Fe $(CN)_6]^{4-/3-}$ being formally counted as organometallic compounds because of the cyanide ligand coordinating via the carbon atom.

In linguistic usage, often no clear distinction is made and both terms – complex and organometallic compound are sometimes used synonymously. Another term frequently used is *metal-organic* – it stands for organic compounds with a metal atom, no matter how the bond between the metal and the organic ligand is realized. Thus it is used alternatively to coordination compounds. A prominent example of this is the term MOF, which stands for metal-organic framework. This refers to compounds that are built up from metal (complex) fragments and organic ligands, i.e. complexes.

The central element of an organometallic compound is the metal-carbon bond. This bond can be more or less polar. It is important to note that the metal is the more electropositive element! Thus $M^{\delta+} - C^{\delta-}$. The polarity of a bond depends on the electronegativity differences of the bonding partners. In Fig. 3.1 the periodic system of the elements with the electronegativities according to the most widely used scales – according to Allred-Rochow and according to Pauling – is given. In addition, it must be taken into

© The Author(s), under exclusive license to Springer-Verlag GmbH, DE, part of Springer Nature 2023
B. Weber, *Coordination Chemistry*,
https://doi.org/10.1007/978-3-662-66441-4_3

electronegativity (EN) Allred-Rochow/Pauling

1	2	3	4	5	6	7	8	9	10	11	12	13	14	15	16	17	18
H 2.20/2.1																	**He** 5.5
Li 0.97/1.0	**Be** 1.47/1.5											**B** 2.01/2.0	**C** 2.50/2.5	**N** 3.07/3.0	**O** 3.50/3.7	**F** 4.10/4.3	**Ne** 4.8
Na 1.01/0.9	**Mg** 1.23/1.2											**Al** 1.47/1.5	**Si** 1.74/1.8	**P** 2.06/2.1	**S** 2.44/2.5	**Cl** 2.83/3.0	**Ar** 3.2
K 0.91/0.8	**Ca** 1.04/1.0	**Sc** 1.20/1.3	**Ti** 1.32/1.5	**V** 1.45/1.6	**Cr** 1.56/1.6	**Mn** 1.60/1.5	**Fe** 1.64/1.8	**Co** 1.70/1.8	**Ni** 1.75/1.8	**Cu** 1.75/1.9	**Zn** 1.66/1.6	**Ga** 1.82/1.6	**Ge** 2.02/1.8	**As** 2.20/2.0	**Se** 2.48/2.4	**Br** 2.74/2.8	**Kr** 2.9
Rb 0.89/0.8	**Sr** 0.99/1.0	**Y** 1.11/1.2	**Zr** 1.22/1.4	**Nb** 1.23/1.6	**Mo** 1.30/1.8	**Tc** 1.36/1.9	**Ru** 1.42/2.2	**Rh** 1.45/2.2	**Pd** 1.3/2.2	**Ag** 1.42/1.9	**Cd** 1.46/1.7	**In** 1.49/1.7	**Sn** 1.72/1.8	**Sb** 1.82/1.9	**Te** 2.01/2.1	**I** 2.21/2.5	**Xe** 2.4
Cs 0.86/0.7	**Ba** 0.97/0.9	**La** 1.08	**Hf** 1.23/1.3	**Ta** 1.33/1.5	**W** 1.40/1.7	**Re** 1.46/1.9	**Os** 1.52/2.2	**Ir** 1.55/2.2	**Pt** 1.42/2.2	**Au** 1.42/2.4	**Hg** 1.44/1.9	**Tl** 1.44/1.8	**Pb** 1.55/1.9	**Bi** 1.67/1.9	**Po** 1.76/2.0	**At** 1.96/2.2	**Rn** 2.1
Fr 0.86/0.7	**Ra** 0.97/0.9	**Ac** 1.00															

Lanthanoide: 1.1 – 1.3 C(sp^3): 2.5 C(sp^2): 2.75 C(sp): 3.29
Actinoide: 1.1 – 1.3

Fig. 3.1 Periodic table of the elements with electronegativities according to Allred-Rochow and according to Pauling. The electronegativities of the hybrid orbitals were calculated according to Mulliken

account that carbon is hybridized differently depending on the ligands. The electronegativity of the carbon is strongly dependent on the degree of hybridization. The reason for this is that the s electrons are subject to a stronger nuclear attraction than the p electrons (with the same principal quantum number). As a result, the electronegativity (EN) of the carbon increases with increasing s-character in the hybrid orbital and, as given in Fig. 3.1, is greatest for an sp hybridized carbon atom. Incidentally, this fact explains why alkanes are so inert. The EN of sp^3 hybridized carbon (2.5) and hydrogen (2.2) are very similar and the C-H bond is very nonpolar. Reagents such as butyllithium (BuLi) or alkylmagnesium halide (Grignard reagent) are commonly used in organic chemistry. These main group organometallic compounds are not discussed in detail in this book. The same applies to other compounds with an element-carbon bond to a nonmetal or semimetal. Such compounds are called element-organic compounds and are also not organometallic compounds in the strict sense.

3.1 History

Formally, the first organometallic compound is Prussian Blue, which was produced by paint manufacturer *Diesbach* in Berlin in the seventeenth century. The complex has a metal-carbon bond, but, like the prussiates of Potash, is not counted among the classical organometallic compounds, but rather among the complexes. The reason for this is that cyanide, as a pseudohalide, is comparable in reactivity to halides. The beginning of organometallic chemistry is associated with the name *Zeise*. *Zeise's* salt $Na[PtCl_3(C_2H_4)]$ was prepared in 1827 and is the first reported olefin complex. Zeise had studied the reaction

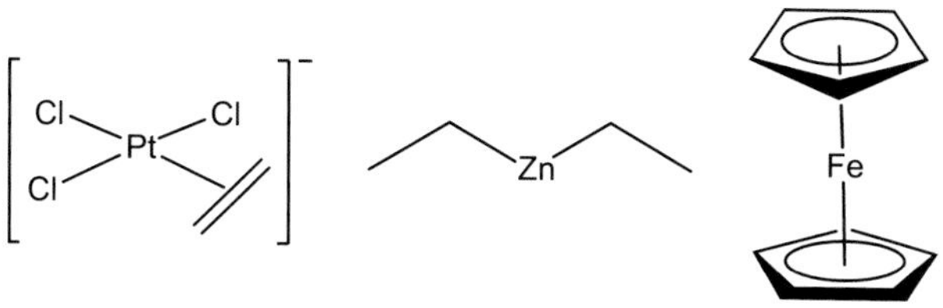

Fig. 3.2 From left to right: Structure of *Zeise's* salt $Na[PtCl_3(C_2H_4)]$, diethylzinc and ferrocene with its sandwich structure

of $PtCl_4$ with ethanol [16]. The complex obtained could later also be prepared by heating the complex $Na_2[PtCl_4]$ in ethanol. In the reaction, the ethanol is dehydrated and ethene is formed, which replaces one of the chlorido ligands at the platinum, resulting in the formation of a metal-carbon bond (Fig. 3.2) [15].

The next compound was diethylzinc (Fig. 3.2) by *Frankland* in 1849, a chance discovery – the aim was to produce the ethyl radical – which was followed by a number of main group metal organyls. The term "organometallic" was coined by *Frankland*. This was followed in 1890 by the aforementioned tetracarbonylnickel(0), the first homoleptic metal carbonyl complex, which is still used today in nickel refining in the *Mond* process named after its discoverer, *L. Mond*. A possibility for C-C bond linkage was developed by *V. Grignard* via the synthesis of alkylmagnesium halides by reacting alkyl halides with magnesium – a discovery for which he was awarded the Nobel Prize in 1912. In 1931, the first hydride complex $[Fe(CO)_4H_2]$ was prepared by *W. Hieber* [17]. In the so-called *Hieber's base reaction*, pentacarbonyliron(0) is converted at basic conditions. In the first step, this produces the ferrate $[Fe(CO)_4]^{2-}$, where the iron has the formal oxidation state -II. This anion is a strong base and accumulates a proton under the reaction conditions, formally increasing the oxidation state of the iron to 0. We obtain the hydride complex $[Fe(CO)_4H]^-$. After acidification, the complex $[Fe(CO)_4H_2]$ [15] is obtained.

In the further course, a large number of different organometallic compounds were produced. A highlight is ferrocene, which was prepared in 1951 by *P. Pauson* and *S. A. Miller*. The discovery of the sandwich structure of this compound in 1953 independently by *E. O. Fischer* and *G. Wilkinson* was recognized with the Nobel Prize in 1973 (Fig. 3.2). Other Nobel Prizes in the field of organometallic chemistry were awarded for technically highly relevant processes such as the preparation of polyolefins from ethylene and propylene (*K. Ziegler* and *G. Natta*, discovery 1955, Nobel Prize 1963) and the hydroboration (*H. C. Brown*, discovery 1956, Nobel Prize 1979). This was followed by a Nobel Prize for the X-ray structural analysis of coenzyme B_{12} (*D. Crowfoot Hodgkin*, discovery 1961, Nobel Prize 1964). The Co-C bond present in the coenzyme is the only known metal-carbon bond in a biologically relevant system that is stable under physiological conditions. The continuing vigorous research in organometallic chemistry (and closely related catalysis) is reflected in the Nobel Prizes of 2001 (*K. B. Sharpless, W. S. Knowles*, and *R. Noyori* for pioneering work in enantioselective catalysis), 2005 (*Y. Chauvin, R. H. Grubbs, R. R.*

Schrock for their work on olefin metathesis), and 2010 (*R. F. Heck, E.-I. Negishi* and *A. Suzuki* for palladium-catalyzed cross-coupling reactions in organic synthesis).

3.2   The 18-Electron Rule

Historically, the noble gas rule was the next model after *Werner's* theory to make a significant contribution to the understanding of chemical bonding in complexes or in organometallic compounds. The noble gas rule (or more modern, 18-electron rule) is based on *Lewis's* octet rule, according to which the main group elements accept or donate electrons until they reach an outer shell of eight electrons, that is, a noble gas shell. *Sidgwick* also explains the coordination compounds with the principle of the noble gas rule. According to this, bonding occurs because each ligand shares a pair of electrons with the central atom, thereby filling up its electron shell until a noble configuration is reached. In the case of organometallic compounds, the 18-electron rule is an excellent way to quickly assess whether a compound is stable. The stability of complexes is explained in a broader context in Chap. 6. Many properties and reactions of organometallic compounds can be explained using the 18-electron rule. In the following, we will look at the basic aspects in more detail.

The difference between the octet rule, where we need (at least for the elements of the first three periods) 8 valence electrons for a noble gas electron configuration, and the 18-electron rule, is the total number of valence electrons. This difference is easily explained by the differences in electron configuration of the metals. In fact, there are also complexes with an 8-electron noble gas configuration. An example of this is the complex $[Al(OH)_4]^-$, which occurs in the *Bayer* process for the preparation of high-purity aluminium, where a noble gas configuration with 8 valence electrons (VE) is obtained. The Al^{3+} has no valence electrons and each OH^- ligand provides two electrons from the free electron pair. This gives the aluminum 8 additional valence electrons, reaching the next noble gas configuration, that of argon. The metal carbonyls are ideal for introducing the 18-electron rule. We consider pentacarbonyliron(0) as an example. This compound is formed from iron and carbon monoxide in an exothermic reaction. In describing the chemical bonding, we had already established that the ligand, as a Lewis base, provides a pair of electrons for the metal center (the Lewis acid) for bonding. The 18-electron rule now assumes that the metal ion uses these electrons to reach noble gas configuration. For this, it needs 18 valence electrons (we already know 8 from the octet rule and 10 are added due to the five *d-orbitals*). The iron is in the oxidation state 0, we determine the number of outer electrons from the electron configuration.

$$\text{Fe} : [\text{Ar}] \, 3d^6 4s^2 \quad \overset{\wedge}{=} \quad 8 \, e^-$$

$$5|C \equiv O| \quad \overset{\wedge}{=} \quad 5 \times 2 \, e^-$$

$$\text{Total} \qquad\qquad\qquad 18 \, e^-$$

Thus, 10 electrons are missing up to 18, which are provided by the five CO ligands. Pentacarbonyleisen(0) is therefore an 18-electron complex and stable. As a further example, we consider tetracarbonylnickel(0), which we know from the *Mond* process.

$$\text{Ni} : [\text{Ar}] \, 3d^8 4s^2 \quad \overset{\wedge}{=} \quad 10 \, e^-$$

$$4|C \equiv O| \quad \overset{\wedge}{=} \quad 4 \times 2 \, e^-$$

$$\text{Total} \qquad\qquad\qquad 18 \, e^-$$

Thus, to reach 18, 8 electrons are missing, which are provided by the four CO ligands. This compound is also a stable 18-electron complex. If we move two elements to the left from the iron, we have chromium. Here we have two d electrons less and to reach a stable 18-electron complex we need 6 carbonyl ligands, which is indeed observed. Between iron and chromium is manganese, which forms a complex with five CO ligands per manganese. If the electrons are counted here ($7 + 5 \times 2 = 17$), the prediction is that this complex cannot be stable because it lacks one electron to a noble gas configuration. This is indeed the case, the complex dimerizes forming a metal-metal bond. In this way, both metal centers share a pair of electrons and each reaches 18 valence electrons.

We consider one last example, namely the yellow and red prussiate of Potash, [Fe(CN)$_6$]$^{4-}$ and [Fe(CN)$_6$]$^{3-}$, and wonder which of the two complexes is more stable. We count the electrons and come to the result:

$$\text{Fe}^{2+}: 3d^6 + 6 \times 2 = 18$$
$$\text{Fe}^{3+}: 3d^5 + 6 \times 2 = 17$$

The iron(II) complex is more stable. In fact, you could eat the compound without getting cyanide poisoning and it is in small quantities an EU-approved salt additive for food (E 535–538, still, please don't try it right away), while the iron(III) complex is a good oxidizing agent and less stable.

On the basis of the 18-electron rule, the existence of a large number of stable organometallic complexes can be predicted. It is only a necessary, but not a sufficient criterion. For f-element organometallic compounds with lanthanides and actinides, a similar approach does not lead to the desired goal. When treating bimetallic complexes as 18-electron complexes with a metal-metal bond, it should be noted that a metal-metal bond predicted in this way does not necessarily mean that this metal-metal bond is actually present. Here, the 18-electron rule quickly reaches its limits.

3.2.1   Counting Electrons

There are several ways to count the electrons that the metal centers and ligands contribute to the total valence electron number of a transition metal complex. One way is to assume that all metal atoms and ligands are in the formal oxidation state zero or charge neutral. The total electron number is obtained by adding the valence electrons of the metal atoms and the electrons contributed by the ligands through the M-L bonds. The resulting electron number is then corrected for the complex charge. For example, a singly bonded Cl atom is a one-electron donor, while in its functions as a μ_2 bridging ligand as a three-electron donor, which – in the sense of the valence bond theory – contributes its unpaired electron to a covalent bond and one of the free electron pairs to a dative-covalent (coordinative) bond.

The second option takes the formal oxidation state or charge into account. Here, in a first step, each ligand is assigned its charge as determined by IUPAC. Then the resulting oxidation state of the metal is determined and subsequently the electron numbers of the central metal and ligands are added. Of the examples considered so far, the calculation procedure is the same for the metal carbonyls because the metal and ligand have the formal oxidation state and charge 0, respectively. For the iron complexes, we have considered the ionic charge of the cyanide ion (-1) as assigned by IUPAC. Had we not done so, we would have had to consider cyanide as a neutral one-electron donor. In this case, the iron would have oxidation state 0 and 8 outer electrons. Then we add one electron per cyanide – we obtain $8 + 6 = 14$ valence electrons for both complexes – and in a last step we have to add the charge of the complex. In the end, the result is the same as for the ionic counting method with 17 and 18 valence electrons for the red (iron(III) complex, three times negatively charged) and the yellow (iron(II) complex, four times negatively charged) complex. At first glance, the ionic counting may seem to make more sense. However, in some complexes the assignment of oxidation states and ionic charges of ligands is difficult, especially when several metal centers are involved. Then the neutral counting becomes more attractive. Ultimately, everyone has to make the decision for themselves!

In the ionic counting method, common neutral molecules such as amines, phosphanes, water, CO, alkenes, etc. are included in the calculation as 2 electron donors; multidentate ligands are treated analogously. NO is a neutral 3 electrondonor according to the IUPAC definition. Since it is often very difficult to determine the charge here, this ligand should always be treated as such. Anions such as hydride, halide, chalcogenide, amide, phosphanide, alkyl, alkylene, alkylidene, etc. contribute 2 electrons per metal center. However, there can be more. At most, there may be as many as free electron pairs are available in suitable spatial alignment; π bonds are usually not counted. Thus, an oxido ligand is 2 electron donor in the terminal binding mode, a 4 electron donor as μ_2-ligand, and a 6 electron donor as μ_3-ligand; the hydrido ligand is 2 electron donor in all binding modes. A η^5-cyclopentadienyl ligand is an anionic 6 electron donor, η^6-benzene is a neutral 6 electron donor. This ionic counting method better reflects the actual chemical bonding and is more consistent with the model ideas. It should be noted that the formal IUPAC oxidation states do not always reflect the actual oxidation states. This is particularly the

Table 3.1 Valency electron number of selected ligands as a function of the binding mode

Ligand	Form of coordination	Valencence electron number	
		Neutral	Ionic
H	μ_1, μ_2, μ_3	1, 1, 1	2, 2, 2
F, Cl, Br, I	μ_1, μ_2, μ_3	1, 3, 5	2, 4, 6
PR_3	μ_1	2	2
CO	μ_1, μ_2, μ_3	2, 2, 2	2, 2, 2
SR/OR	μ_1, μ_2, μ_3	1, 3, 5	2, 4, 6

case for so-called "*non-innocent ligands*". In Table 3.1, important valence electron numbers are once again summarized.

Two Examples for Electron Counting
In principle, everyone must decide for themselves which counting method to use. In this book, the ionic counting method is used, unless explicitly stated otherwise. In the following examples, it is demonstrated that the same results are obtained with both variants.

Tetracarbonyldiiodidoiron(II), [Fe(CO)$_4$I$_2$]

$$
\begin{array}{llll}
\text{neutral :} & & \text{ionic :} & \\
\text{iron}(0) & 8 & \text{iron}(2+) & 6 \\
4\ \text{CO} & 4 \times 2 & 4\ \text{CO} & 4 \times 2 \\
2\ \text{I}(0) & 2 \times 1 & 2\ \text{I}(-1) & 2 \times 2 \\
\text{total} & 18 & \text{total} & 18
\end{array}
$$

Tris(μ-bromido)hexacarbonyldimanganate(1-), [(CO)$_3$Mn(μ-Br)$_3$Mn(CO)$_3$]$^-$

$$
\begin{array}{llll}
\text{neutral :} & & \text{ionic :} & \\
2\ \text{magnanese}(0) & 2 \times 7 = 14 & 2\ \text{magnanese}(1+) & 2 \times 6 = 12 \\
6\ \text{CO} & 6 \times 2 = 12 & 6\ \text{CO} & 6 \times 2 = 12 \\
3\ \mu_2\ \text{Br}(0) & 3 \times 3 = 9 & 3\ \mu_2\ \text{Br}(1-) & 3 \times 4 = 12 \\
\text{Charge } []^{-1} & 1 & & \\
\text{total} & 36 & \text{total} & 36
\end{array}
$$

It should be noted again that we determine the valence electron number at the metal center. This usually corresponds to the number of d-electrons, but it may be that, based on the electron configuration of the atom, s-electrons are still present, which must then be taken into account. This is the case, for example, with metals in the formal oxidation state

0. In compounds, all valence electrons are first assigned to the d orbitals, which in this case are energetically below the s orbitals. The iron(0) in a compound is thus not a $3d^6$ $4s^2$ element, as it is usual for the electron configuration of the atom, but a $3d^8$ element. For all the complexes listed in this section, the 18-electron rule applies. For many other complexes it plays no role.

3.3   Important Reactions in Organometallic Chemistry

The listing of the historical highlights of organometallic chemistry has already shown that many achievements are closely linked to catalytic processes, which are still widely used in large-scale processes today and have enormous relevance for society. All these catalytic processes can be divided into a few basic reactions that are characteristic of organometallic chemistry. Knowledge of these basic reactions helps in understanding catalytic processes. For this reason, the five most important reactions are presented below. It should be noted from the outset that all reactions are equilibrium reactions. The basic reactions can be recognized by a characteristic change in the valence electron number (VE) of the metal ion, the coordination number (CN) and the formal oxidation state (OS) of the metal. In the following equations, [M] represents a complex fragment at which the reaction takes place. The first reaction is a very general reaction, which is also important for complexes.

3.3.1   Coordination and Release of Ligands

When a ligand is coordinated to a metal or complex fragment (Fig. 3.3), the coordination number (CN) of the metal increases by one and the valence electron number (VE) by two, because the new ligand provides two more electrons for the metal center. A common type of reaction is ligand substitution. The pathway of this reaction consist of the release and addition of a ligand. The sequence can also be reversed or synchronized, within one elementary step. In Fig. 3.4 the three possible mechanisms are given.

In mechanism (a), the rate-determining step is the release of a ligand, and the coordination of the new ligand occurs rapidly. This mechanism is called dissociative substitution mechanism. In variant (b), first the new ligand L_B is bound to the metal center by increasing the coordination number and then ligand L_A is released. This mechanism, in which the rate-determining step is the attachment of the new ligand, is called an associative substitution mechanism. The third variant (c) is a synchronous mechanism, in which the attachment of L_B and cleavage of L_A occur simultaneously. Which mechanism takes place depends strongly on the complex fragment (increase in coordination number possible or not). Square planar complexes with 16 valence electrons often prefer the associative mechanism, because here the coordination number can be increased easily. For octahedral complexes, especially with 18 valence electrons, the dissociative mechanism rather takes place.

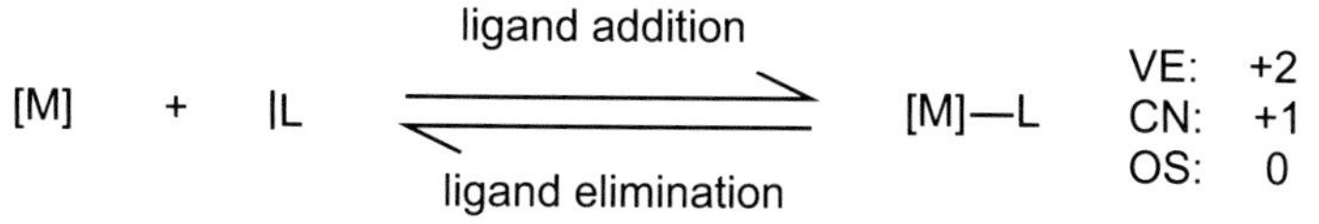

$$[M] \quad + \quad IL \quad \underset{\text{ligand elimination}}{\overset{\text{ligand addition}}{\rightleftharpoons}} \quad [M]-L \qquad \begin{array}{ll} \text{VE:} & +2 \\ \text{CN:} & +1 \\ \text{OS:} & 0 \end{array}$$

Fig. 3.3 Coordination and release of a ligand to/from a complex fragment

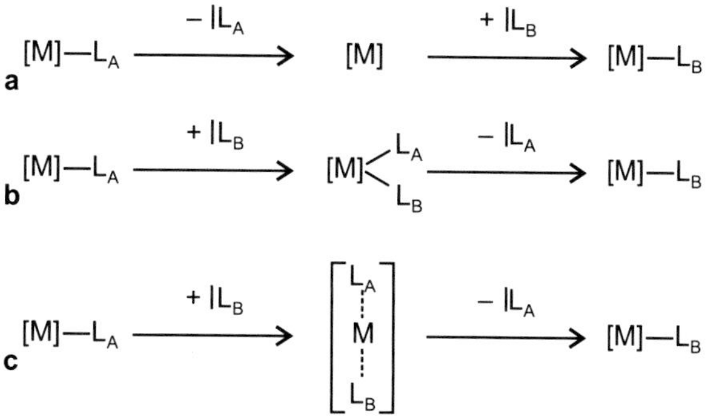

Fig. 3.4 The ligand substitution reactions can proceed according to different mechanisms: (**a**) dissociative, (**b**) associative and (**c**) synchronous (analogous to the S_N2 mechanism from organic chemistry)

$$\textbf{a} \quad [M]-L_A \xrightarrow{-\,IL_A} [M] \xrightarrow{+\,IL_B} [M]-L_B$$

$$\textbf{b} \quad [M]-L_A \xrightarrow{+\,IL_B} [M]{<}^{L_A}_{L_B} \xrightarrow{-\,IL_A} [M]-L_B$$

$$\textbf{c} \quad [M]-L_A \xrightarrow{+\,IL_B} \left[\begin{array}{c} L_A \\ M \\ L_B \end{array} \right] \xrightarrow{-\,IL_A} [M]-L_B$$

3.3.2   Oxidative Addition and Reductive Elimination

In the oxidative addition of a compound A-B to a metal centre (Fig. 3.5), the oxidation state at the metal centre increases by two and the coordination number also increases by two. Accordingly, metal centers in low oxidation states with a low coordination number are suitable for this reaction. They are very common in d^8 complexes (M = Pd^{II}, Pt^{II}, Rh^I, Ir^I) and d^{10} complexes (M = Pd^0, Pt^0, Au^I). It must be possible to increase the oxidation state by two for metals suitable for this reaction. Complexes suitable for oxidative addition have less than 18 valence electrons and have free coordination sites. The reaction involves a change from a square-planar coordination geometry (CN 4) to an octahedral one (CN 6) or from a linear/angled one to a square-planar one. Typical substrates A-B for the addition reaction are compounds such as H_2, X_2 (X = Cl, Br, I), HX and halohydrocarbons RX (R = alkyl, aryl, vinyl, alkynyl, ...). In the case of hydrocarbons and silanes, oxidative addition can lead to the activation of C-H or Si-H bonds, and possibly also to C-C or Si-Si bond activation. Typical products of reductive elimination are H_2, X_2, HX and RX. This corresponds to the reversal of the oxidative addition reaction described before. For reductive elimination, the substituents must be *cis to* each other. Reductive elimination involving C-C bond formation is an important step in a variety of catalytic reactions.

3.3.3   Insertion of Olefins and β-H Elimination

The insertion of olefins into a metal-hydrogen or a metal-carbon bond is an important step in many catalytic processes, such as polymerization catalysis. In a first step, an olefin

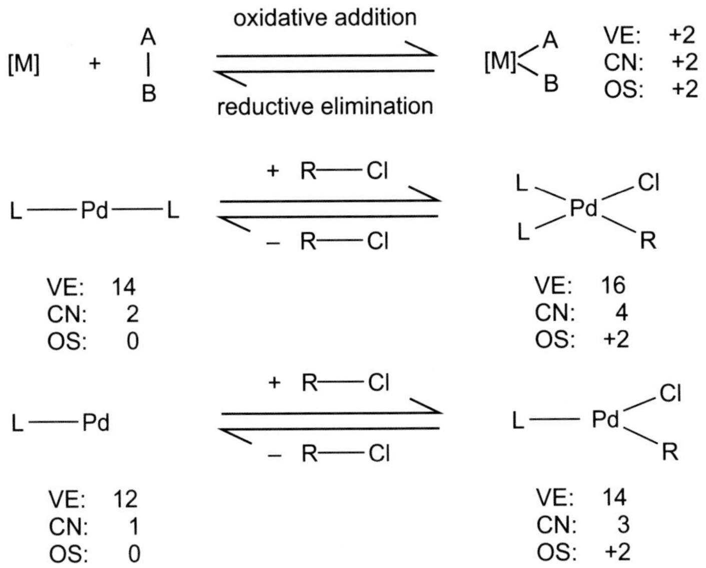

oxidative addition

$$[M] \quad + \quad \begin{array}{c} A \\ | \\ B \end{array} \quad \underset{\text{reductive elimination}}{\overset{\text{}}{\rightleftharpoons}} \quad [M]{<}\begin{array}{c} A \\ B \end{array} \qquad \begin{array}{ll} \text{VE:} & +2 \\ \text{CN:} & +2 \\ \text{OS:} & +2 \end{array}$$

$$L{-\!\!-}Pd{-\!\!-}L \quad \underset{-\ R{-\!\!-}Cl}{\overset{+\ R{-\!\!-}Cl}{\rightleftharpoons}} \quad \begin{array}{c} L \\ \\ L \end{array}{>}Pd{<}\begin{array}{c} Cl \\ \\ R \end{array}$$

VE:	14		VE:	16
CN:	2		CN:	4
OS:	0		OS:	+2

$$L{-\!\!-}Pd \quad \underset{-\ R{-\!\!-}Cl}{\overset{+\ R{-\!\!-}Cl}{\rightleftharpoons}} \quad L{-\!\!-}Pd{<}\begin{array}{c} Cl \\ \\ R \end{array}$$

VE:	12		VE:	14
CN:	1		CN:	3
OS:	0		OS:	+2

Fig. 3.5 Oxidative addition and reductive elimination

coordinates to a free coordination site of the metal (addition of the ligand), which is inserted into the M-H or M-C bond in a second step. This again generates a free coordination site to which a new olefin can coordinate. The insertion occurs via a four-membered transition state. When an olefin is inserted into an M-C or M-H bond, the oxidation state of the metal center remains constant, the coordination number decreases by one, and the valence electron number decreases by two, because there is one less ligand bound to the metal center in the product. The reverse reaction is the β-H elimination (β-hydride elimination), in which a hydride is split off to form a C-C double bond.

Agostic interactions between hydrogen at the β carbon and the metal ion are a prerequisite for the elimination of β hydrides. Agostic interactions refer to binding interactions between a Lewis acid atom and bonding electrons in complexes. In the β-hydride elimination, these are the C-H bonding electrons. A four-membered, cyclic, non-planar transition state follows, leading to the formation of the C-C double bond. Figure 3.6 shows the mechanism of β-H elimination or olefin insertion.

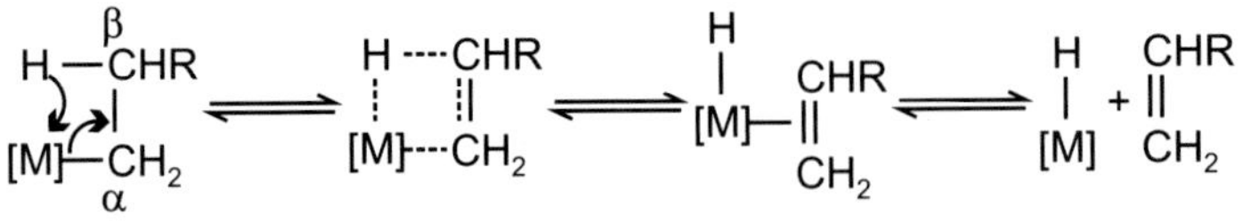

$$\begin{array}{c} \overset{\beta}{H}{-}CHR \\ {[M]}{-}CH_2 \\ \scriptstyle\alpha \end{array} \rightleftharpoons \begin{array}{c} H\cdots CHR \\ \vdots \quad \| \\ {[M]}\cdots CH_2 \end{array} \rightleftharpoons \begin{array}{c} H \\ | \quad CHR \\ {[M]}{-}\| \\ CH_2 \end{array} \rightleftharpoons \begin{array}{c} H \quad CHR \\ | \ + \ \| \\ {[M]} \quad CH_2 \end{array}$$

Fig. 3.6 Mechanism of β-H elimination as well as the reverse reaction, olefin insertion. The cyclic transition state is not planar

In some catalysis cycles, β eliminations, which are usually reversible, are to be suppressed. The reaction mechanism shows that this can be achieved by ligands that do not carry a β standing hydrogen atom. A second possibility is a coordinatively saturated metal center that has no possibility of forming an agostic interaction. A third possibility is that the formation of the leaving alkene is sterically or energetically unfavorable. The mechanism of β-H elimination has been demonstrated, among others, by labeling experiments with deuterium (deuterium at β carbon is found in the elimination product).

In addition to the 1,2-insertion shown, a 1,1-insertions can also take place.

3.3.4 Oxidative Coupling and Reductive Cleavage

In oxidative coupling, two alkenes or alkynes are coupled to form a C-C bond (Fig. 3.7). As with oxidative addition, this reaction increases the formal oxidation state by two, but the coordination number does not change. The reaction is relevant for ethylene oligomerizations.

3.3.5 α-H Elimination and Carbene Insertion

In some cases, a α standing hydrogen can be eliminated instead of the β standing hydrogen. The corresponding reaction is the α-H elimination (Fig. 3.8). This reaction generates a

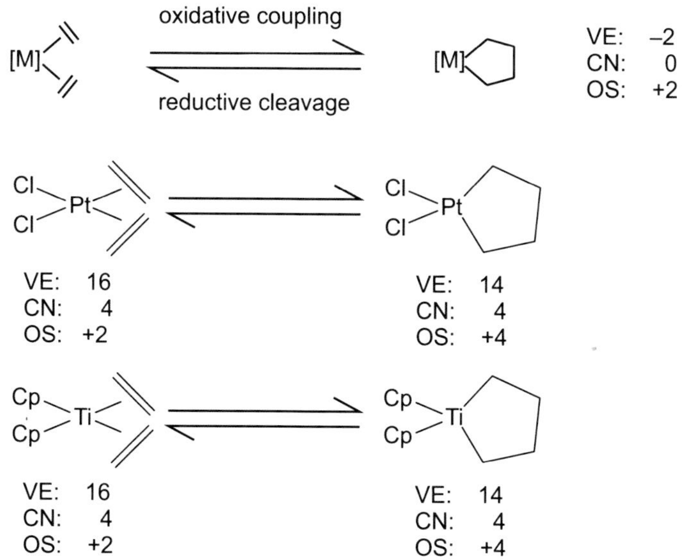

Fig. 3.7 Oxidative coupling and reductive cleavage

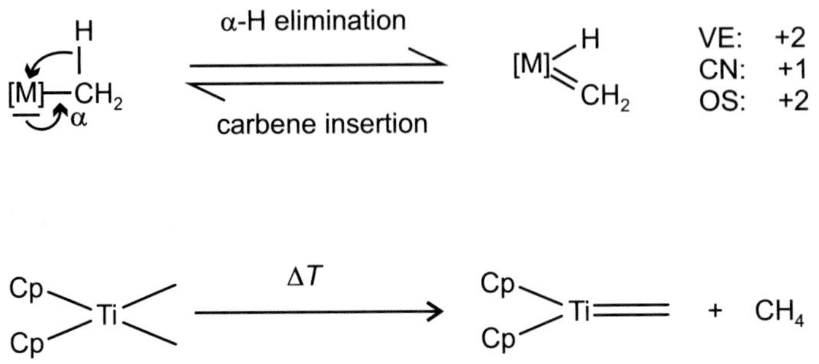

Fig. 3.8 α-H-elimination and carbene insertion

carbene attached to the metal center, which is used, for example, in olefin metathesis. The α-H elimination is observed, among others, when no β standing hydrogen is present. The carbene is often observed only as an intermediate, is difficult to isolate, and is very reactive. The reaction is reversible and the reverse reaction is carbene insertion.

3.4   Questions

1. Discuss the stability of the complexes mentioned in Chap. 2 in items 1–4 with the help of the 18-electron rule (give the calculation method)!
2. Which of the complexes from Chaps. 1 and 2 belong to the organometallic compounds according to IUPAC?
3. What conditions must a metal center meet for oxidative addition/reductive elimination to occur?
4. Name the five basic reactions of organometallic chemistry.

Bond Theories

4

In the following, one must always keep in mind that a model is only a model and not reality. In a model, various assumptions and simplifications are used to explain central properties of a system. Reality is arbitrarily complicated! We will examine each model in terms of its explanatory potential for the various properties of complexes, pointing out its strengths and its limitations. Already explained was the 18-electron rule, which can be used to make statements about the coordination number of organometallic compounds. In the following, the valence bond theory (VB-), the ligand field theory (LF-) and the molecular orbital theory (MO-) are presented in the order of their historical development. Complex properties requiring explanation include coordination number (CN), coordination geometry, stability of complexes, their color, and magnetism. For the models discussed below, it is necessary to know or determine the valence electron configuration of the metal centers. For this reason, we begin with a review of the essentials of electron configuration and term symbols, which we need for an understanding of the further chapters.

4.1 Electron Configuration and Term Symbols

4.1.1 Quantum Numbers

Our idea for an atom is that it is made up of a positively charged nucleus around which the negatively charged electrons move. The shape, expansion and energy of the space occupied by the electron, which we call the *orbital,* is determined by the quantum numbers. The first quantum number is the principal quantum number n, which describes the energy and spatial extent of the orbitals. As the principal quantum number n increases (an integer value; $n = 1$, 2, 3, ...), the energy of the orbital increases. The second quantum number is the angular

Fig. 4.1 Energy level diagram
of the atomic orbitals defined by
principal and angular
momentum quantum number
according to the Aufbau
principle

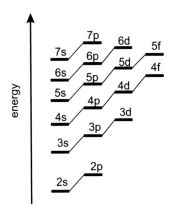

momentum quantum number l, which is also called the azimuthal quantum number. It takes integer values in the range from 0 to $n–1$. This means that for $n = 2$, angular momentum quantum numbers with values 0 and 1 are obtained. The angular momentum quantum number describes the spatial shape of the orbitals, more precisely the number of nodal planes. At a nodal plane, the sign of the wave function describing the orbital changes. As the number of nodal planes increases, the energy of the orbitals increases, but not to the same extent as for the increase in the principal quantum number, n. The angular momentum quantum number is denoted by letters. For $l = 0, 1, 2, 3, \ldots$ the letters s, p, d, f (further in alphabetical order) are used. The 2 s orbital is an orbital with $n = 2$ and $l = 0$, while for the 3p orbital $n = 3$ and $l = 1$. The energetic order of the orbitals determined by the principal and angular momentum quantum numbers is given in Fig. 4.1. The third quantum number is the magnetic quantum number m_l. It describes the orientation of the orbitals in space. The numerical values go from $l, l - 1, \ldots, 0, \ldots, - l + 1, - l$; for a angular momentumg quantum number l there are$2l + 1$ magnetic quantum numbers. Consequently, for $l = 1$ there are three magnetic quantum numbers with values 1, 0 and $- 1$. These are called p_x-, p_y- and p_z-orbitals. Orbitals with the same principal and angular momentum quantum numbers and different magnetic quantum numbers have the same energy. The orbitals are then said to be degenerate. In the presence of an external magnetic field (in the z-direction), this degeneracy is removed and the electrons in orbitals with different m_l have different potential energies. This led to the name *magnetic* quantum number. The first three quantum numbers describe the orbitals, i.e. the quantum state of an electron in an atom. Figure 4.2 shows the spatial shape and orientation of the individual orbitals. Each orbital can be occupied by a maximum of two electrons. According to their energetic position, the energetically lower orbitals are occupied first. The necessity of a fourth quantum numbers was first recognized by *Pauli*. The *Pauli exclusion principle,* named after him, requires that two electrons in an atom must differ in at least one quantum number. In other words; No two electrons in an atom are identical in all four quantum numbers. This fact is ensured by the fourth quantum number, the spin quantum number m_s. This quantum number describes

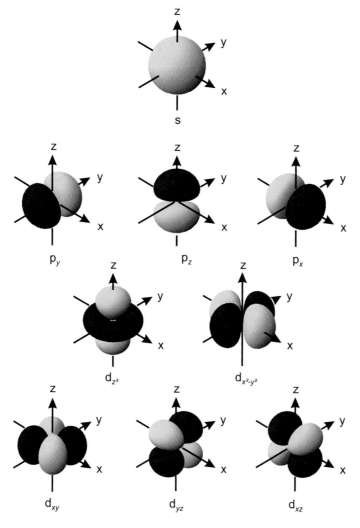

Fig. 4.2 Representation of the 1s, 2p and 3d orbitals. A nodal plane is formed when the sign changes between the orbital lobes. We see that the 1s orbital with the angular momentum quantum number 0 has no nodal plane, the 2p orbitals with the angular momentum quantum number 1 have one nodal plane, and the 3d orbitals with the angular momentum quantum number 2 have two nodal planes [18]

the orientation of the electron spin s and can take the value $+\frac{1}{2}$ or $-\frac{1}{2}$. To symbolize this, the electron is represented as an arrow pointing up ($\uparrow$, spin up) or down ($\downarrow$, spin down). The electron belongs to the group of particles with half-integer spin called fermions. All fermions obey Pauli's exclusion principle.

The electron configuration of an element indicates how the electrons are distributed among the individual orbitals. The number of electrons (and protons) is determined by the atomic number of the element, e.g. ^{11}Na or ^{26}Fe. These electrons are then used to fill the orbitals according to the energy level scheme following the *Aufbau* principle, thus starting with the lowest available energy. It should be noted that an s orbital can occupy two electrons, the three degenerate p orbitals six, the five d orbitals 10 and the seven f orbitals 14. When the d and f orbitals are occupied, there are sometimes deviations in the electron configuration from the *Aufbau* principle due to the rule that fully and half-occupied shells are particularly stable. In the following, the electron configurations of sodium, iron and chromium are given as examples (we restrict ourselves to metals). We see that in chromium the 4s orbital, which according to the energy level scheme is first filled with two electrons, is only singly occupied. Instead, the 3d shell is half occupied with five electrons and this electron configuration is more stable (energetically lower) than the possibility with two electrons in the 4s orbital and four electrons in the five 3d orbitals.

Na (11) $1s^2\, 2s^2\, 2p^6\, 3s^1 = $ [Ne] $3s^1$
Fe (26) $1s^2\, 2s^2\, 2p^6\, 3s^2\, 3p^6\, 3d^6\, 4s^2 = $ [Ar] $3d^6\, 4s^2$
Cr (24) $1s^2\, 2s^2\, 2p^6\, 3s^2\, 3p^6\, 3d^5\, 4s^1 = $ [Ar] $3d^5\, 4s^1$

The valence electrons are the electrons in the outermost shells after the last complete noble gas configuration (core electrons). Fully filled shells are always particularly stable. That is why the noble gases, as the name already says, are very noble, i.e. not very reactive. The valence electrons are more easily removed and are important for bond formation and the chemical reactivity in general. All reactions, no matter if they lead to the formation of salts, molecules or complexes, have the goal to achieve a fully filled noble gas shell for the atoms involved (see 18-electron rule or chemical bonding). The rule that fully occupied shells are particularly stable can also be applied in a weakened form to half-occupied shells. This helps in determining important oxidation states for the elements – an important step when we consider complexes. Here we often want to know how many d electrons the metal center has. In our examples, the most important oxidation state for sodium is +I, where it has noble gas configuration. For iron, the most important oxidation state (where it is prevalent in naturally occurring compounds) is +III.

At this point, an important digression on the occupation of the 3d and 4s orbitals is made, since it is precisely here that complications can arise. If one follows the *Aufbau* principle sketched in Fig. 4.1, which still goes back to *N. Bohr*, one finds an apparently preferential occupation of 4s before 3d. This sequence has no theoretical basis, but is purely phenomenological. It stems from *Bohr*'s analysis of atomic spectra, which clearly shows so-called s states. If we now generate the chemically relevant and familiar metal ions (e.g. the divalent cations M^{2+}), spectroscopy again clearly points to d states.

Metal: $d^{n+2}\, s^0 \leftrightarrow \mathbf{d^n\, s^2}$ (with exceptions, e.g. Cu, Cr)
M^{2+}: $\mathbf{d^n\, s^0} \leftrightarrow d^{n-2}\, s^2$ (without exceptions)

So there is an apparent contradiction here, in that the occupation with electrons and the release of electrons both take place preferentially in the s orbitals. To clarify this, it should first be said that Bohr's principle of structure correctly reflects the final configurations of the metal atoms, but does not allow any statement about the order of occupation. This is shown by the example of scandium, for which one finds:

Sc(III): [Ar] $3d^0 4s^0$
Sc(II): [Ar] $3d^1 4s^0$
Sc(I): [Ar] $3d^1 4s^1$
Sc(0): [Ar] $3d^1 4s^2$

Orbital energies are variable and depend on the number of electrons present while the nuclear charge remains constant. In fact, during the transition Sc(III)$\rightarrow$ Sc(II), a 3d orbital is actually occupied first and only afterwards two electrons move into the 4s orbital. In many textbooks, the following (not entirely correct) explanation can be found: as soon as the first electron occupies the 3d level, it is "lowered" and is energetically below the full 4s level in the occupied state. The more electrons there are in the 3d level, the greater the energy difference becomes. This explains the experimental finding that in ions the 4s orbitals are not occupied. Also the fact that especially for the late 3d elements (when the 3d level is already relatively full and the energetic distance to the 4s level is large) the oxidation state +II is the most important one (Co, Ni, Cu, Zn) can be explained vividly in this way. In the case of the iron(III) ion, an electron is also removed from the 3d shell. By this it is half-filled and particularly stable. Important oxidation states of chromium are +III and + VI. Only the oxidation state +IV (in chromates) can be explained well with the electron configuration. In this case the noble gas configuration is reached again. To explain the oxidation state +III, we need the ligand field theory.

4.1.2   Term Symbols

There are cases where the electron configuration alone is not meaningful enough to explain observed phenomena. When degenerate orbitals are partially occupied, there are different ways to distribute the electrons among the orbitals leading to different (total) angular momentum quantum numbers. Fully filled shells have no contribution (first Hund's rule). We consider as an example the titanium(III) ion. It has a d^1 electron configuration. This signifies that the only valence electron occupies one of the d orbitals. The question that now arises is which of the orbitals is occupied. Since we have five d orbitals, there are five possible electron configurations. Each of these states is called a microstate. Since the five d orbitals all have the same energy, the energy of the five different microstates is also the same. The microstates are degenerate. Microstates of the same energy are combined into one term, which is represented by a term symbol. To find out which microstates of an element belong to the ground state and which belong to excited states, we need Hund's rules.

Hund's Rules

1. The ground state of a system is the state in which as many electrons as possible have a parallel spin. Or: Degenerate orbitals are filled up in such a way that they are initially simply occupied with electrons of the same spin. In short, this means for the ground state that the total spin S of the system should be maximal (maximum multiplicity).
2. The ground state of a system with a given multiplicity is the state where the total orbital angular momentum quantum number L is maximal. This means that degenerate orbitals are filled in such a way that first those with a high magnetic orbital angular momentum quantum number are occupied.
3. The ground state of a system with given S and L is the state in which, if the shell is less than half-filled, the total angular momentum quantum number J *is* the lowest, or, if the shell is more than half-filled, the total angular momentum quantum number J is the highest.

The total spin quantum number S as well as the total orbital angular momentum quantum number L of a system can be easily determined by summing the spin quantum numbers or orbital angular momentum quantum numbers of the individual electrons. $S = \sum m_s$ and $L = \sum m_l$. The angular momentum j describes the spin-orbit coupling between the intrinsic angular momentum and the orbital angular momentum of the electron. The total angular momentum J is here calculated from L and S with all values between $L + S$ and $L - S$.

Determination of Term Symbols

The term symbol $^M L_J$ consists of a capital letter for the total orbital angular momentum L. For $L = 0, 1, 2, 3, \ldots$. the letters S, P, D, F, . . .(continue in alphabetical order) are used. They are the same letters as for the orbital angular momentum l, i.e. the orbitals, only this time they are capitalized.

In the upper left corner the multiplicity M of the term is given. We calculate it from the total spin S according to the formula $M = 2S + 1$. It should be briefly noted that S is used twice – once as the total spin S written in italics and for $L = 0$ written as "normal" S.

The total angular momentum J is indicated on the bottom right corner.

As a first example, we consider the sodium atom in the ground state and excited state. In the ground state, there is an unpaired electron in the 3s orbital. The total spin quantum number $S = s = \frac{1}{2}$, the total orbital angular momentum quantum number $L = l = 0$ and for the total angular momentum quantum number J there is only one possible numerical value, namely $\frac{1}{2}$. For the term symbol, we calculate the multiplicity $M = 2$ from the total spin S. Now we have all the information and can write down the term symbol. It is $^2S_{\frac{1}{2}}$. In the first excited state, the valence electron is in the energetically next higher orbital, the 3p orbital. The total spin or multiplicity does not change. Since the unpaired electron is now in a p orbital, it can assume the orbital angular momentum quantum number 1, 0, or -1. We apply Hund's second rule to this excited state. It tells us that L should be maximal. Therefore, for the determination of the term symbol, we assume that the electron is in the orbital with $l = 1$. Thus, the total orbital angular momentum quantum number is $L = 1$. This implies that this state is $2L + 1 = 3$-fold degenerate. This is consistent with our previous assumptions. We had said that the three p orbitals all have the same energy. So it should not matter in which of the three orbitals the electron is. In the absence of an external magnetic field, the three possible microstates have the same energy and belong to the same term symbol. If an external magnetic field is applied, Hund's second rule applies and the state at which L is maximal is the ground state. The total angular momentum quantum number can now take two possible values: $L + S = \frac{3}{2}$ and $L - S = \frac{1}{2}$. The third of Hund's rules tells us that when the shell is less than half-filled (which is the case here), J *is* minimal in the ground state. Please note that Hund's rules can also be applied to excited states. The term symbol of the first excited state is $^2P_{\frac{1}{2}}$ and energetically slightly above it is the second excited state $^2P_{\frac{3}{2}}$. The energetic difference between the two excited states is not large, which is why both transitions are excited and why we observe two lines in the sodium emmision spectrum, for example. At this point it should be pointed out that Hund's rules are used to determine the ground state and are not always suitable for excited states.

As a second example, we consider the iron(III) ion in the ground state. The five valence electrons are distributed among the five d orbitals. The total spin $S = 5 \times \frac{1}{2} = \frac{5}{2}$. Thus we obtain as multiplicity $M = 6$. The total orbital angular momentum $L = 2 + 1 + 0 + -1 + -2 = 0$. There is only one possible total angular momentum with the value $J = \frac{5}{2}$. The term symbol is $^6S_{\frac{5}{2}}$. As a final example, we consider a chromium(III) ion. Here we have 3d electrons distributed among the five orbitals, and there are again several ways to do this. In determining the ground state, Hund's rules again help us; S and L should be maximal. We obtain as values $S = \frac{3}{2}$, $M = 4$ and $L = 3$. The state is sevenfold degenerate with respect to orbital angular momentum, meaning that there are seven ways to distribute the three electrons with preserved spin among the five orbitals, all of which have the same energy. The total angular momentum can take the values $J = \frac{9}{2}, \frac{7}{2}, \frac{5}{2}$ and $\frac{3}{2}$. Since the 3d shell is less than half filled, the term symbol for the ground state is $^4F_{\frac{3}{2}}$.

Throughout this book, we will frequently return to terms and their associated term symbols.

4.2   The Valence-Bond (VB) Theory

The valence bond approach to bonding in complexes uses similar assumptions as the 18-electron rule: The ligands each provide two electrons to the metal ion to fill the electron shells. Unlike the 18-electron rule, the electrons are not simply added together, but the empty orbitals of the metal center are filled up. Before filling up the orbitals, one forms the required number of hybrid orbitals from the s, p and d orbitals. We already know the principle of hybridization from carbon chemistry. From orbitals of different energy, "hybrid orbitals" of the same energy are created, which also have the same spatial shape. This model is used in carbon to construct identical C-H bonds for methane, CH_4, even though the carbon atom has the valence electron configuration $2s^2\ 2p^2$. Four sp^3 hybrid orbitals are created, which are needed for the four bonds. In the case of carbon, this is done by first lifting an electron from the s orbital into the last empty p orbital before the hybrid orbitals are formed. The VB theory shows parallels to organic chemistry and consequently is best applied when the bonds are predominantly covalent. It is not applicable to complexes where the bonds are predominantly ionic. In contrast to organic chemistry, it does not use half-occupied orbitals to create hybrid orbitals, but empty ones. Before the required number of hybrid orbitals are formed, the electrons in the 3d orbitals can be paired to create empty orbitals for hybridization. Alternatively, spin pairing can be omitted and a portion of the 4d orbitals used instead. This gives rise to the so-called inner-shell (spin pairing) and outer-shell (no spin pairing) complexes. The now more common variants to denote the two different spin states are low-spin and high-spin – these terms come from ligand field theory. The nature of the hybrid orbitals can be used to some extent to make statements about the coordination geometry. Six d^2sp^3 orbitals form an octahedron and four sp^3 orbitals form a tetrahedron just as in carbon. In contrast, four dsp^2 orbitals are responsible for a square planar coordination environment. A major drawback of the VB model is that it does not give any predictions for the color of the complexes, nor does it explain why spin pairing occurs in some cases and not in others. For this reason, it is hardly used today and is also mentioned in this book only for the sake of completeness. As examples, the electron configurations for an iron(III) complex with six ligands (Fig. 4.3) and a nickel(II) complex with four ligands (Fig. 4.4) are presented below in the box notation used with this model. For a given number of ligands and known oxidation state, the structure and magnetism of the complexes can be correlated.

The starting point is the electron configuration of the metal centre. We consider as an example an iron(III) complex, the electron configuration of iron(0) is [Ar] $3d^6\ 4s^2$ and for the trivalent oxidation state [Ar] $3d^5$. If we assume that there is an outer-shell complex, the next step is to fill the empty orbitals with the electrons from the ligands. Starting from six ligands, 12 electrons have to be distributed. The energetically lowest lying orbitals are filled up first. This means two electrons occupy the 4s orbital, six electrons occupy the three 4p orbitals, and the last four electrons occupy two 4d orbitals. The concept of hybridization is used to obtain equivalent orbitals for ligand coordination. Six sp^3d^2 hybrid orbitals are formed, yielding an octahedron. When an inner-shell complex is formed, the five electrons

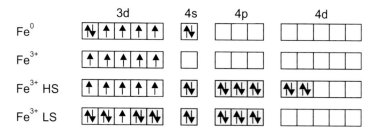

Fig. 4.3 From top to bottom in box notation: of iron(0), electron configuration from iron(III) ion, electron configuration of an iron(III) outer-shell complex with six ligands (six sp^3d^2 hybrid orbitals, octahedron), and electron configuration of an iron(III) inner-shell complex with six ligands (six d^2sp^3 hybrid orbitals, octahedron). Hybridized orbitals are marked in grey, for further explanation see text

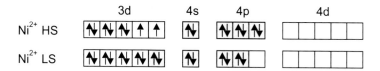

Fig. 4.4 From top to bottom in box notation: Electron configuration of a nickel(II) outer-shell complex with four ligands (four sp^3 hybrid orbitals, tetrahedral) and electron configuration of a nickel (II) inner-shell complex with four ligands (four dsp^2 hybrid orbitals, square planar). Hybridized orbitals are marked in grey, for further explanation see text

in the 3d orbitals are paired first. This creates two empty 3d orbitals, which are filled up together with the 4s and 4p orbitals. The two 4d orbitals are not needed for this. Six d^2sp^3 hybrid orbitals are formed, which also form an octahedron.

The nickel(II) complex has four ligands. This time, the 4s orbital and the three 4p orbitals are required for the outer-shell complex. Four sp^3 hybrid orbitals are formed for the eight electrons of the ligands and the coordination environment is a tetrahedron, analogous to carbon. Assuming an inner-shell complex, the electrons in the 3d orbitals are paired and an empty 3d orbital is available. In addition, the 4s and two of the 4p orbitals are still needed to form the four dsp^2 hybrid orbitals. This combination results in a square planar coordination environment for the nickel. Indeed, nickel(II) complexes are diamagnetic when they have a square-planar coordination environment. All electrons are paired. If the coordination environment is tetrahedral or octahedral, a paramagnetic complex is obtained.

Compared to the 18-electron rule, the VB theory is an advance. It can be used to establish a correlation between structure and magnetism. However, it does not explain why in some cases an outer-shell complex is formed and in other cases an inner-shell complex. It fails in explaining the color of coordination compounds. The ligand field theory, which follows in the historical sequence, remedies these two deficiencies and has therefore prevailed over the VB method. In the case of the VB method, the progress compared to the 18-electron rule is so small (if one holds the ligand field theory against it) that it has only historical significance and is no longer used.

4.3   The Ligand Field Theory

The best way to begin the introduction to ligand field theory is to throw everything we have learned so far overboard. The ligand field theory is a further development of the crystal field theory for transition metal salts. For this theory, only electrostatic interactions are considered. In ligand field theory, this consideration is extended to include additional parameters (e.g., the Racah parameter). We thus move away from the notion of the Lewis acid-base adduct and the donation of electron pairs from the ligands and consider only the electrostatic interactions between the positively charged central atom, the negatively charged ligands, and the negatively charged electrons in the d orbitals. Neutral ligands such as water or ammonia are considered as dipole molecules that are negatively polarized toward the metal center. For simplicity, we assume that the ligands and the central atom are represented as point charges (negatively and positively charged). This assumption is at the same time a weakness of the model, because it does not allow it to be applied to compounds such as metal carbonyls, where both the metal center and the ligand are uncharged. Nevertheless, we will find that this model is ideally suited to satisfactorily explain the properties of the complexes listed further above (in particular, color, structure, and magnetism). The properties of the metal carbonyls, in particular their stability, can in many cases already be explained by the 18-electron rule.

Since each ligand corresponds to a point charge and thus generates an electric field, the fields of ligands bound to the metal centre overlap and form a common ligand field. Due to the interaction of the central ion with the ligand field, the outer electrons, i.e., the (3)d electrons, are particularly affected. It should be noted, however, that to simplify the ligand field theory, the ligands are initially assumed to have point charges, which means that electronic effects within the electron shell of the ligands are initially ignored. The ligands (negative point charges) are arranged around the metal center (positive point charge) in such a way that they are as far apart as possible. With this we have – except for the square planar coordination geometry, which we will come to later – already explained most of the coordination geometries. The number and arrangement (polyhedra, distance) of the ligands results in the so-called *ligand field*. The equilibrium distance between metal center and ligands is based on the attractive (different charge) and repulsive (same charge of the negatively charged electrons in the d-orbitals and the ligand point charges) interactions. This distance depends on the metal center (oxidation state, position in the periodic table) and the ligands (charge or strength). Due to the different spatial orientation of the d orbitals, the repulsive interactions between the ligands and the electrons in the orbitals are different. To understand this better, let us take another closer look at the structure of the d orbitals in Fig. 4.2.

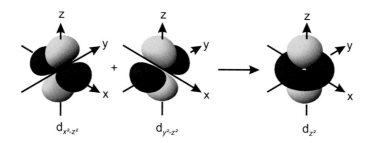

Fig. 4.5 Derivation of the shape of a d_{z^2} orbital

The first thing we notice is that the d_{z^2} orbital differs in structure from the other four orbitals. The other four orbitals each consist of four orbital lobes, two with a positive sign and two with a negative sign, always arranged alternately in a plane, so that two nodal planes are obtained in each case. The orbital name provides information about the position of the nodal planes (and thus also about the position of the orbital lobes). In the d_{xy} orbital, the nodal planes lie in the xz and the yz planes, and the orbital lobes lie between them in the xy plane. We observe the same principle for d_{xz} and d_{yz} orbital. The $d_{x^2-y^2}$ orbital is rotated $45°$ about the z-axis compared to the d_{xy} orbital. The orbital lobes are still in the xy-plane, but here they are exactly on the x and y axis and the nodal planes are in between. The d_{z^2} orbital seems to be out of frame. There are two orbital lobes on the z axis (in analogy to the $d_{x^2-y^2}$ orbital), but the two nodal planes are not planes but cone- or funnel-shaped along the z axis. Instead of the other two orbital lobes, we have a ring in the xy plane. The reason for this is that in analogy to the $d_{x^2-y^2}$ orbital we could formulate two more orbitals, which follow in their shape the first one, namely the $d_{x^2-z^2}$ and the $d_{y^2-z^2}$ orbital. Since we need only five d orbitals due to the quantum numbers, a linear combination of the two remaining variants is used for the last orbital. A correct name for this would be $d_{2z^2-x^2-y^2}$ orbital, which will be abbreviated as d_{z^2} for simplicity. In Fig. 4.5 we see that the shape is now explained conclusively.

4.3.1   Octahedral Ligand Field

Having clarified this issue, we can return to the repulsive interactions between the electrons in d orbitals and the ligands. As a first example, we consider an octahedral ligand field. Six ligands approach the metal centre in our imagination along the three axes of the coordinate system, as given in Fig. 4.6. In Fig. 4.7, the energy of the d orbitals is given as a function of the environment. For the free ion, the five orbitals are degenerate. When the ion is placed in an initially spherically symmetric (spherical) ligand field, the 3d orbitals are raised in energy due to the repulsive interaction between the electrons in the d orbitals and the negative charge of the ligand field. We use this energy level as a reference for all further considerations. In Fig. 4.6, we see that the distance between the orbital lobes of the d

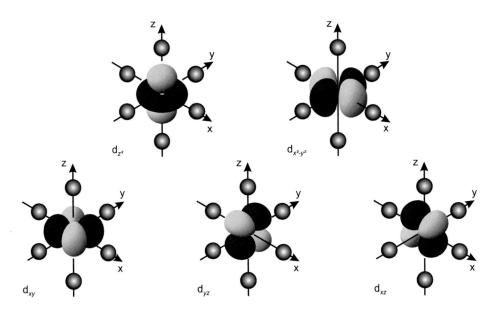

Fig. 4.6 Interaction of the d orbitals with the six ligands of an octahedral ligand field

orbitals and the ligands varies at fixed equilibrium distance, leading to different strengths of repulsive interactions. In the octahedral ligand field, the repulsion is particularly large for the orbitals whose orbital lobes lie on the axes (d_{z^2} and $d_{x^2-y^2}$ orbital). These orbitals are referred to as the e_g orbitals. For the other three orbitals (d_{xy}, d_{xz} and d_{yz} orbital) the repulsion is considerably smaller. These orbitals are called t_{2g} orbitals. Here the orbital lobes point between the axes of the coordinate system and the interaction with the negative point charges of our ligands is weaker. This leads to an energetic splitting of the orbitals into orbitals that are energetically favorable for the electrons (d_{xy}, d_{xz} and d_{yz} orbitals, here the repulsive interactions for the electrons are not so large) and orbitals that are energetically unfavorable (d_{z^2} and $d_{x^2-y^2}$). This energetic splitting of the orbitals in the ligand field is the central element of the ligand field theory. The nature of the splitting (which orbitals are favorable and which are unfavorable) is controlled by the coordination geometry. The strength of the splitting (energy difference between the d orbitals) is influenced by the ligands and the metal center. In the octahedral ligand field, the splitting is denoted as Δ_O or 10 Dq. The two e_g orbitals are raised by 6 Dq, while the three t_{2g} orbitals are lowered by -4 Dq each. If all orbitals are equally occupied (single, double, or empty), then there is no energy gain for the system.

We summarize again the process given in Fig. 4.7. A central ion with field-free and degenerate d orbitals (Fig. 4.7 (1)) is placed in a spherical (= spherically symmetric) ligand field. The electrons in the d orbitals experience a repulsive force from the ligand field and the energy of all orbitals is raised by the energy ε_0 compared to the field-free state (Fig. 4.7 (2)). Because of the spherical symmetry of the field, no splitting occurs and the orbitals

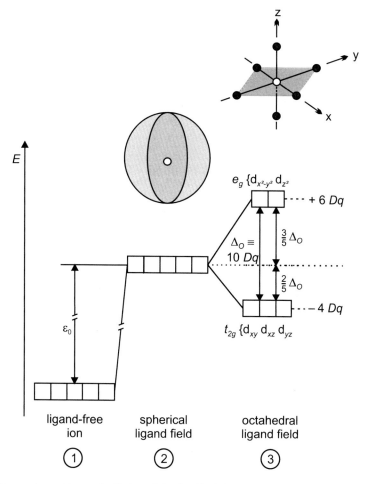

Fig. 4.7 Energetic position and splitting of the d orbitals in the octahedral ligand field

remain degenerate. However, if the ligands form a field that is generated in a particular coordination polyhedron and is therefore no longer spherically symmetric, the d orbitals undergo energetic splitting depending on their orientation in the field. In octahedral complexes (Fig. 4.7 (3)) [ML_6], the ligands converge along the three axes of the coordinate system. Orbitals whose orbital lobes also lie on the axes and thus point to the ligands ($d_{x^2-y^2}$, d_{z^2}) experience a stronger repulsion and are energetically raised, whereas orbitals which lie between the axes (or ligands) (d_{xy}, d_{xz}, d_{yz}) are energetically lowered. Thus, the geometrical considerations show: in the octahedral ligand field, the degeneracy of the d orbitals is partially cancelled, with the $d_{x^2-y^2}$ and d_{z^2} orbitals being energetically raised (e_g orbitals) and the d_{xy}, d_{xz} and d_{yz} orbitals (t_{2g} orbitals) being energetically lowered. The magnitude of the splitting of the orbitals in the octahedral ligand field, denoted Δ_O or

10 *Dq*, depends on the strength of the ligand field and is on the order of 7000–40,000 cm^{-1}. However, the ratio of raising and lowering is independent of the ligand field and is an amount weighted according to the centroid theorem, since one orbital can only be lowered (raised) by the amount of energy by which another has been raised (lowered). Thus, the ratio is precisely fixed according to the law of the center of gravity. So, starting with the five d orbitals, this means that the energetic raising of the two e_g orbitals is $\frac{3}{5}\Delta_O$ or 6 *Dq*, *and the* lowering of the three t_{2g} orbitals is $\frac{2}{5}\Delta_O$ or 4 *Dq*.

4.3.2 Crystal Field Stabilization Energy and the Spectrochemical Series

Due to the splitting of the d-orbitals in the ligand field and the fact that the energetically lower lying orbitals are occupied first, there is an energy gain for a complex compared to the same system in the spherical ligand field with degenerate d orbitals. The energy gain is characteristic for the respective complex and depends on the central ion, the ligands and their geometrical arrangement, i.e. the ligand field (or more precice crystal field since only point charges are considered). The term *crystal field stabilization energy, CFSE,* has been defined to describe this effect quantitatively:

> The crystal field stabilization energy (CFSE) is defined as the energy by which a system split in the ligand field is more stable than ε_0. Here ε_0 is the energy of the electrons in the field before the splitting of the d orbitals, i.e. in the spherical ligand field with degenerate d orbitals.

For example, for a d^1 central ion ($=$ d^1 spin system) in an octahedral ligand field, the d electron will occupy one of the energetically lower t_{2g} orbitals. The configuration is referred to as $(t_{2g})^1$. The energy in such a system relative to the unsplit state corresponds to -4 *Dq* – the ligand field stabilization energy of a d^1 system. For a d^2 and d^3 system, the amount increases to -8 and -12 *Dq*, respectively. Above a number of 4d electrons, there are two variants of how the electrons can be distributed among the d orbitals. In Fig. 4.8 both variants are compared. In the high spin case, all orbitals are occupied individually and the CFSE corresponds to $3(-4) + 6 = -6$ *Dq*. In the low spin case, all four electrons are placed in t_{2g} orbitals and the CFSE corresponds to -16 *Dq*. Here we have to consider that the spin pairing energy *P* has to be applied, which is subtracted from the CFSE. The spin pairing energy is the energy that must be applied when an orbital is occupied by a second electron. A simple idea would be that repulsive interactions between the two electrons are responsible for this. The question of which of the two possibilities occurs depends on the magnitude of the splitting of the d orbitals.

The magnitude of the splitting (and hence the coloration and electron spectrum of the complexes, see Color of Complexes) depends on the nature of the ligands and the central ion. The following trends were found in systematic comparative studies:

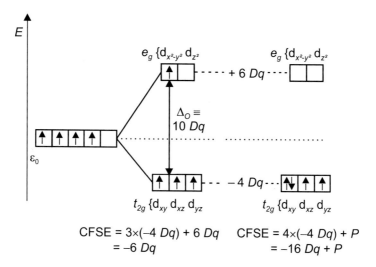

$$CFSE = 3\times(-4\ Dq) + 6\ Dq \qquad CFSE = 4\times(-4\ Dq) + P$$
$$= -6\ Dq \qquad\qquad = -16\ Dq + P$$

Fig. 4.8 CFSE for a d^4 ion in the octahedral ligand field with and without spin pairing

- For the same ligands, the Dq values increase with the oxidation state of the central ion.
- With the same ligands, the Dq values increase by 30–40% when moving from one transition metal series to the other.
- Complexes with central ions of the same transition metal series show similar Dq values at the same oxidation state and with the same ligands.

Sorting the Dq values by increasing values for complexes of the same central ion with different ligands gives the following *spectrochemical series*:

$$I^- < Br^- < SCN^- < Cl^- < N_3^- \approx F^- < OH^- < O^{2-} < H_2O < NCS^- < NH_3$$

$$\approx C_5H_5N\ (py) < NH_2CH_2CH_2NH_2\ (en) < bipy \approx phen < NO_2^- < H^- < CN^-$$

$$< PR_3 < CO$$

A similar series as for the ligands exists also for the metal ions. If the complexes $[ZL_6]$ with the same ligands are ordered according to increasing Dq values, the following *spectrochemical series* of the metal ions is obtained:

$$Mn^{2+} < Ni^{2+} < Co^{2+} < V^{2+} < Fe^{3+} < Cr^{3+} < Co^{3+} < Ru^{2+} < Mn^{4+} < Mo^{3+} < Rh^{3+}$$

$$< Ir^{3+} < Re^{4+} < Pt^{4+}$$

The magnitude of the ligand field splitting $10\ Dq$ depends on the metal ion and the ligands. It can be estimated from the product of a factor f_L, which depends only on the ligand, and a factor g_M, which depends only on the metal center. Some values are summarized in Table 4.1 [3].

Table 4.1 g_M-values of selected metal ions and f_L-values of selected ligands (for six ligands) [3]

M	$g_M/1000\ cm^{-1}$	L	f_L
Mn^{2+}	8.5	Br^-	0.72
Ni^{2+}	8.9	SCN^-	0.73
Co^{2+}	9.3	Cl^-	0.80
Fe^{3+}	14.0	F^-	0.90
Cr^{3+}	17.0	ox^{2-}	0.99
Co^{3+}	19.0	H_2O	1.00
Ru^{2+}	20.0	py	1.23
Rh^{3+}	27.0	NH_3	1.25
Ir^{3+}	32.0	en	1.28
		bipy	1.33
		CN^-	1.70

$$10Dq = g_M \times f_L$$

We see that the values reflect the *spectrochemical series* of the ligands and the metal ions. As an example, we estimate the ligand field splitting for a hexacyanidochromat(III) ion and a hexafluoridochromat(III) ion. As numerical values we obtain $17 \times 1.7 \times 1000\ cm^{-1} = 28{,}900\ cm^{-1}$ for the cyanido complex and $17 \times 0.9 \times 1000\ cm^{-1} = 15{,}300\ cm^{-1}$ for the fluorido complex. The spectroscopically determined values for Δ are $26{,}600\ cm^{-1}$ and $15{,}060\ cm^{-1}$. We see that for both complexes the calculated value agrees quite well with the spectroscopically determined value. The limitations of this simple estimation are shown when we consider the hexacyanidoferrate(III) ion. Here there is a large discrepancy between the calculated ($14 \times 1.7 \times 1000\ cm^{-1} = 23{,}800\ cm^{-1}$) and experimentally determined ($32{,}200\ cm^{-1}$) value. This is due to the fact that different spin states are possible in octahedral iron(III) complexes, which were not considered here.

4.3.3   High-Spin and Low-Spin

The energetic splitting of the d orbitals allows us to explain the magnetic properties of the complexes in more detail. We again take an iron(III) complex with six ligands. A high-spin complex is obtained when the splitting of the d orbitals in the ligand field Δ_O is significantly smaller than the spin-pairing energy P (the energy that must be applied when two electrons must share an orbital). Conversely, a low-spin complex is obtained when Δ_O is significantly larger than P. In this case, it is energetically more favorable to pair the electrons in the energetically lower orbitals. Since the spin pairing energy P is always of the same order of magnitude, the observed spin state depends primarily on the size of the splitting of the ligand field. As already mentioned, this is influenced by the ligands and the metal center. An iron(III) complex with six water molecules as ligands is in the high-spin state, while the same metal center with six cyanide ions is a low-spin complex. The hexaaquacobalt(II) ion is a high-spin complex, while the hexaaquacobalt(III) ion is a low-spin complex.

4.3.4   Non-Octahedral Ligand Fields

Square Planar Ligand Field   In VB theory, it has already been mentioned that d^8 complexes are often square planar and diamagnetic. Examples are $[Ni(CN)_4]^{2-}$, $[PtCl_2(NH_3)_2]$ or $[AuCl_4]^-$. To explain this issue, we need to look at the splitting of the d orbitals in the square planar ligand field. This can be derived from the octahedral ligand field by assuming that the ligands are removed in the z direction. This leads to a stabilization of all orbitals with z component, while the others are destabilized by the corresponding weighted amount. As a result, the d_{z^2} orbital is strongly lowered in energy compared to the octahedral field, while the orbitals lying in the xy plane, i.e. $d_{x^2-y^2}$ and d_{xy}, are raised in energy. The orbitals d_{xz} and d_{yz} are only slightly lowered. The centroid theorem has to be considered again. As can be seen in Fig. 4.9, this leads to a very pronounced destabilization of the $d_{x^2-y^2}$ orbital, while the other four orbitals are relatively close together. Which order the energetically lower orbitals take depends strongly on the ligands. They decide whether, for example, the d_{z^2} orbital is energetically above or below the d_{xy} orbital. The exact order of the orbitals is not relevant for the high stability of square planar d^8 systems. It is due to the pronounced destabilization of the $d_{x^2-y^2}$ orbital, which is not occupied in the low-spin case (the complexes are always diamagnetic). The other four orbitals are all fully occupied and this leads to a particularly high ligand field stabilization energy. This explains why quadratic planar d^8 complexes are always diamagnetic.

The **distorted octahedron** is the transition from the ideal octahedron to the square planar ligand field. For an octahedron stretched along the z axis, the splitting of the d orbitals lies between that in the octahedral and that in the square planar ligand field, as shown in Fig. 4.9. For an octahedron compressed along the z axis, the orbitals with z-portion are raised and the orbitals which lie in the xy plane are lowered. The **square pyramidal ligand field** can be constructed by removing only one ligand along the z axis. If we assume that the ligands are always the same, then the splitting of the d orbitals lies between that in the stretched octahedron and that for the square planar complex (see Fig. 4.9).

Tetrahedral and Cubic Ligand Field   In the tetrahedral ligand field, the ligands approach the metal center exactly between the axes – the opposite case to the octahedron. Accordingly, the orbitals d_{xy}, d_{xz} and d_{yz} are raised in energy, while the orbitals $d_{x^2-y^2}$ and d_{z^2} are lowered. Since fewer ligands are involved in the formation of the ligand field, if the ligands are the same, the splitting size ratios are:

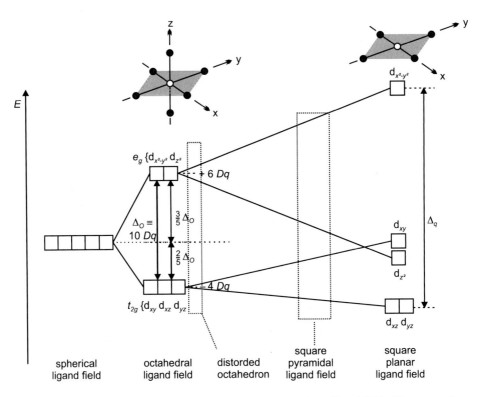

Fig. 4.9 Splitting of the d orbitals in the octahedral and square planar ligand fields. The square planar ligand field emerges from the octahedral one by removing the two ligands along the z-axis. As a result, the orbitals with z component decrease in energy and the orbitals in the x-y plane are raised in energy. For the distorted octahedron (stretched along the z axis) and a square pyramidal ligand field (only one ligand on the z axis is removed), the splitting of the d orbitals can be generated analogously. The splitting in these two cases is not as large as in the case of the square planar coordination environment

$$\Delta_T = \frac{4}{9}\Delta_O$$

As shown in Fig. 4.10, a cube can be composed of two tetrahedra. Accordingly, in a cube-shaped (cubic) ligand field, instead of four ligands, eight ligands approach the metal center between the axes. The splitting of the d orbitals follows the same rules, but is twice as large compared to the tetrahedron with the same ligands. Figure 4.11 summarizes once again the different splittings discussed.

Fig. 4.10 Composition of a
cube from two tetrahedra

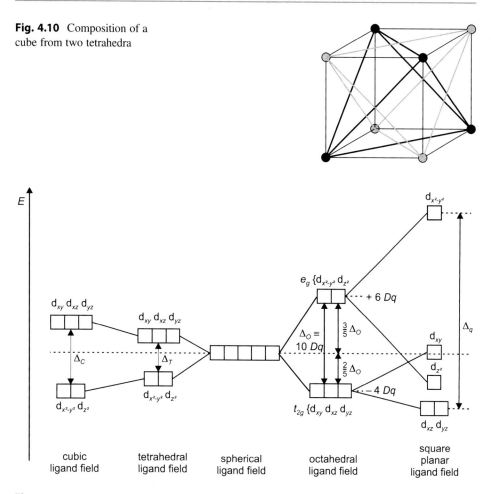

Fig. 4.11 Energetic position and splitting of the d orbitals as a function of the ligand field with the assumption that ligands and central atom remain the same

$$\Delta_C = 2\Delta_T = \frac{8}{9}\Delta_O$$

4.4   The Molecular Orbital (MO) Theory

For some classes of complexes, e.g. carbonyls and their derivatives, olefin, N_2 or sandwich complexes, the ligand field theory has proved to be unsuitable for interpreting the stability and observed properties of the complexes. In this case, molecular orbital theory (MO theory for short) has proved to be very useful. MO theory states that atomic orbitals overlap

Fig. 4.12 MO diagram for the interaction between two s-orbitals using the example of the hydrogen molecule. A binding σ and an antibinding σ^* molecule orbital is formed

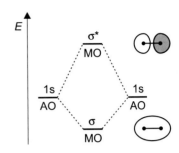

to form bonding and antibonding molecular orbitals. The binding molecular orbital is energetically more favorable than the two atomic orbitals, while the antibonding molecular orbital is energetically less favorable. If each atomic orbital is only singly occupied, both electrons can be accommodated in the energetically favorable molecular orbital, resulting in an energy gain for the system. A prerequisite for overlap is that the atomic orbitals have suitable symmetry. As an example to get you started, the molecular orbital scheme (MO scheme for short) for the hydrogen molecule is shown in Fig. 4.12.

If the charge density of the molecular orbital is on the core bond axis, as in the example of H_2, it is a σ bond; if the charge density is above and below the core bond axis (which corresponds to a nodal plane), it is a π bond. This can occur when p orbitals are involved in bond formation, as shown in Fig. 4.13 for the NO radical example.

This concept can now be transferred to coordination compounds. We now assume that the orbitals of the ligand interact with the d orbitals of the metal center. Here it is again important to consider the symmetry of the orbitals. Generally, a distinction is made between σ complexes and π complexes. In the case of σ complexes, only σ bonds are present between the metal center and the ligands, whereas in the case of π complexes, π bonds are also formed, which can have a decisive influence on the properties. Before we draw up the MO scheme for a σ and a π complex, we will take a closer look at the σ and π interactions between metal and ligand.

4.4.1 σ and π Interactions Between Ligand and Central Atom

So far, we have considered the bond between metal and ligand as an electrostatic interaction (ligand field theory) or as a Lewis acid-base adduct, where the shared electron pair is provided by the ligand. Transferring the latter approach to molecular orbital theory, this corresponds to a σ-donor-acceptor bond between a pair of electrons in a nonbonding (donor) orbital of the ligand and an empty acceptor orbital of the central atom. However, pure metal-ligand σ bonding rarely occurs and additional ligand-metal π interactions can significantly affect the properties of the complexes. Ligand-metal π interactions are distinguished between π donors and π acceptors. A good σ and π donor ability have hard alkoxido- and also amidoligands, which are well suited to stabilize early transition metals

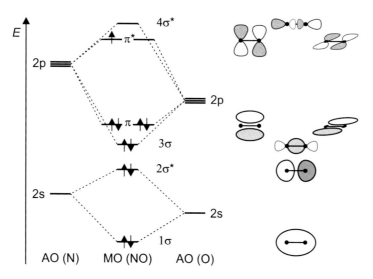

Fig. 4.13 MO scheme for the NO radical *(left)* with the associated molecular orbitals *(right)*. The 1 s level has not been shown for clarity. Not to scale!

in high oxidation states. These have (almost) no d electrons and the empty d orbitals can serve well as acceptor orbitals for the full donor orbitals of the ligand. These are usually fully occupied p orbitals. If the metal-ligand bond lies along the z axis of the coordinate system, the p_z orbital has a suitable symmetry for the formation of a σ bond, e.g. with the d_{z^2} orbital. The p_x and p_y orbitals on the ligand are now available for the formation of π bonds, as shown in Fig. 4.14. The corresponding binding partners at the metal ion would be the d_{xz} and the d_{yz} orbital. In the case of alkoxido and amido ligands, only one π bond is usually formed. Even more pronounced are π donor bonds between metal and ligand in complexes with imido or oxido ligands, in which the ligands can form up to two π bonds to the central atom in addition to the σ-donor bond. In summary, we note that a prerequisite for pronounced π donor interactions are empty d orbitals at the metal center, which interact with the occupied orbitals of the ligand. This is usually realized in early transition metals in higher oxidation states, which are stabilized by halido, alkoxido or oxido ligands.

A second variant of π interactions occurs when the metal center is an electron-rich (usually late) transition metal in low oxidation states. Here, occupied d orbitals of the metal centre can act as donors to the ligand. The prerequisite for this is that empty orbitals in the suitable energy range and with the correct symmetry for the formation of a π bond are present on the ligand. In this case, the bond is referred to as a π back bond, which means that the ligand is a π acceptor. This type of bond is often observed with metal carbonyls. In the CO ligand, energetically relatively low-lying empty π^* orbitals are present, which have a suitable symmetry to overlap with fully occupied d orbitals of the metal center (d_{xz}, d_{yz} or d_{xy}). Although CO behaves only as a weak Lewis base, it nevertheless forms extremely stable complexes with transition metals at low oxidation states. This finding cannot be

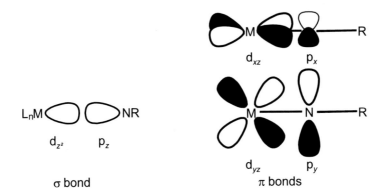

Fig. 4.14 Simplified representation of the molecular orbitals for the σ and π bonds of an imido-metal bond. The formation of both π bonds in the presence of suitable acceptor orbitals at the metal centre leads to an approximate linearity of the R–N = M unit

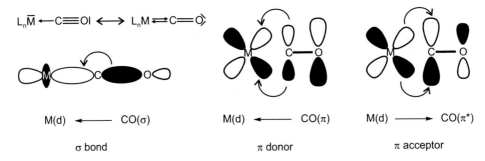

Fig. 4.15 Simplified representation of the molecular orbitals of the σ and π bonds of a carbonyl-metal bond and the corresponding mesomeric boundary structures. The π donor interaction shown in the middle is usually negligible

attributed solely to the properties of the metal-ligand σ bond. Rather, this is an example of a metal-ligand bond whose stability is largely determined by a π interaction. The corresponding interactions are shown schematically in Fig. 4.15.

The largest contribution to the strength of the metal-ligand bond is the σ donor interaction of the energetically highest occupied σ orbital of the carbonyl ligand with acceptor d orbital of the metal center. This bonding increases the electron densityon the metal center. This in turn is partially returned to the ligand by the π bonding. This occurs through the overlap of an occupied d orbital of suitable symmetry with an empty π* orbital of the carbon monoxide. The "backflow" of electron density to the ligand via the π bond then in turn again increases the σ donor capability of the donor orbital on the carbon atom. The σ donor and π backbonding influence each other in the sense of strengthening the metal-ligand bond, which is why one speaks of a σ-donor-π-acceptor synergism. Theoretically, π donor bonding between the occupied π orbitals of CO and empty d orbitals at the metal

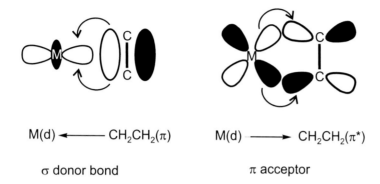

$$M(d) \longleftarrow CH_2CH_2(\pi)$$ $$M(d) \longrightarrow CH_2CH_2(\pi^*)$$

σ donor bond   π acceptor

Fig. 4.16 Simplified representation of the molecular orbitals of the σ and π bonds of an ethene-metal bond. The occupied binding π orbital of ethene is responsible for the σ donor properties and the empty antibinding π^* orbital acts as π acceptor

center is also possible for the carbonyl ligand. This plays a minor role for π-acidic ligands such as CO (the acidic here refers to Lewis acidity). Due to the π-backbonding, the formal bond order of the CO bond is lowered by the (partial) occupation of antibonding orbitals. In valence structural formulas, this circumstance can be represented by a mesomeric boundary structure with a double bond between metal and carbon and a double bond between C and O, as shown in Fig. 4.15.

The π-backbonding also plays a major role in the coordination of olefins to metal centres, such as in Zeis salt (Sect. 3.1). Figure 4.16 shows the σ and π interactions that occur in this process. In this case, the binding π orbital of ethene is responsible for the σ donor bond.

4.4.2   MO Scheme of a σ Complex

We have seen that in complexes, just as in molecular compounds, σ and π bonds can be formed. The MO theory for complexes is analogous to that for molecular compounds. The orbitals of the ligands and the metal overlap to form bonding and antibonding molecular orbitals, with the bonding ones being energetically lower and occupied by electrons first. This leads to an energy gain which is responsible for the stability of the complex. The better the orbitals overlap, the stronger the splitting of the molecular orbitals and the greater the energy gain.

For simplicity, we consider an octahedral 3d complex. Each of the six ligands provides a fully occupied orbital of suitable symmetry for the formation of a σ bond. This can be an s or p orbital of suitable symmetry, but also a molecular orbital. The six σ ligand orbitals combine to form six ligand group orbitals of suitable symmetry to form bonding and antibonding interactions with the valence orbitals of the metal center. In the octahedral complex, the bonds from the metal to the six ligands lie on the axes of the coordinate

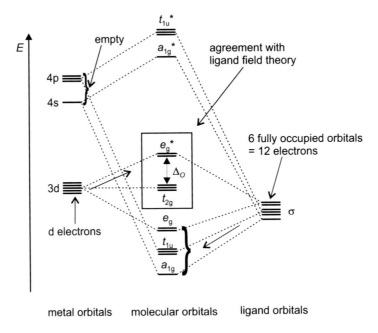

Fig. 4.17 MO scheme of an octahedral σ complex. The six orbitals of the ligands are fully occupied. The twelve electrons fill up the six lowest energy orbitals (a_{1g}, t_{1u} and e_g) of the complex. The electrons from the d orbitals of the metal are now distributed between the t_{2g} and e_g^* orbitals. In accordance with the ligand field theory, the splitting between the orbitals determines whether Hund's rule (small splitting Δ_O, high-spin complex) or Pauli's principle (large splitting Δ_O, low-spin complex) is applied

system. Not all valence orbitals of the metal center are suitable for the formation of a σ bond to the ligand. Of the partially occupied 3d orbitals, only the d_{z^2} and the $d_{x^2-y^2}$ orbitals are suitable for this purpose, in which the orbital lobes lie on the axes. The other three d orbitals (d_{xz}, d_{yz} and d_{xy}) do not have the appropriate symmetry. They are said to be non-bonding and are present in the complex unchanged with the same energy, as seen in Fig. 4.17. The empty 4s and the three empty 4p orbitals also possess suitable symmetry to form a σ bond. The formation of six bonding and six antibonding molecular orbitals occurs as given in Fig. 4.17. On the part of the ligands, a total of twelve electrons are made available for the molecular orbital scheme. These formally occupy the six binding molecular orbitals a_{1g}, t_{1u} and e_g. On the metal side, the 3d orbitals are partially occupied, and the 4s and 4p orbitals are empty. These electrons are distributed among the three nonbonding 3d orbitals and the two antibonding e_g^* orbitals that emerged from the d_{z^2} and $d_{x^2-y^2}$ orbitals. The energetic splitting decides whether Hund's rule is followed in the process and we obtain a high-spin complex, or the electrons are paired in the lower-lying orbitals. These five molecular orbitals correspond to the t_{2g} and e_g^* orbitals that we already know from ligand field theory. They are in fact nonbonding and antibonding, respectively. Here both models overlap, the same result is obtained for the distribution of the d electrons. The molecular orbital theory additionally explains to us why the t_{2g} orbitals obtained also with the ligand field theory are nonbonding and the e_g orbitals are antibonding. No term symbols are used for the designation of the orbitals, but a scheme that we find again later in the splitting term symbols (see Tanabe-Sugano diagrams). The letters a, e, and t stand for singly, doubly, and triply degenerate states, respectively. That is, one, two, or three orbitals of equal energy are present in the MO scheme of the complex, respectively. The letters g and u stand for german *gerade* (*even*) and *ungerade* (*odd*) and denote the parity of the orbital (the behavior of the orbital under point reflection, see selection rules for electronic transitions).

4.4.3 MO Scheme of a π Complex

In complexes containing π bonds, the previously nonbonding t_{2g} orbitals participate in π bonds with the ligand. For this, three further ligand orbitals are needed, which have a suitable symmetry. We have already discussed which ones are suitable for this purpose. The three d orbitals d_{xz}, d_{yz} and d_{xy} of the metal can overlap with p or π orbitals of the ligands of suitable symmetry to form bonds with π symmetry. In the full MO scheme, the five 3d orbitals are subdivided into two e_g orbitals (d_{z^2} and $d_{x^2-y^2}$) and three t_{2g} orbitals (d_{xz}, d_{yz} and d_{xy}), similarly to the σ complexes. For the ligand orbitals, we distinguish whether the π orbitals are energetically above or below the σ orbitals. In the first case we have empty π acceptor orbitals available for backdonation, while in the second case the ligand orbitals are occupied and they are π donor ligands. Figure 4.18 shows the influence of the position of the ligand p or π orbitals on the splitting between the t_{2g} and the e_g orbitals. We see that for π donor ligands the splitting becomes smaller and for π acceptor ligands the splitting Δ_O is larger than in the system without π bonds. This allows us to explain why CO is a strong-field ligand and fluoride is a weak-field ligand at the other end of the spectrochemical series. For both extreme cases, substantial double bond contributions can be formulated via π interactions. Here it is precisely the π acceptor or π donor character that determines the direction. Ammonia and pyridines belong to the ligands without clear π effects. They are in the middle of the spectrochemical series.

4.5 Questions

1. What is a ligand field, who generates it, and who does it affect?
2. Calculate the CFSE for a d^1 and d^4 system in the octahedral and tetrahedral ligand field.
3. $FeCr_2O_4$ crystallizes in the spinel structure, $CoFe_2O_4$ in the inverse spinel structure. What are the differences? Explain the facts with the help of the crystal field theory (CFSE)!
4. With which of the following ligands would you expect a complex of the general composition $[Fe^{II}L_6]$ to be a high-spin or a low-spin complex? H_2O, CN^-, F^-, SCN^-, NH_3, bipy ($= 2$ L).

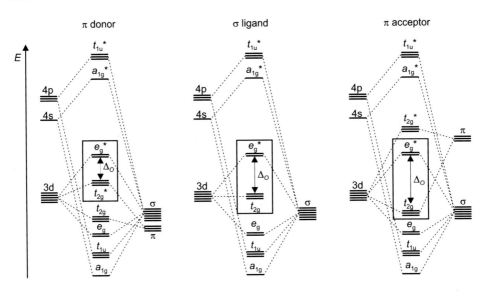

Fig. 4.18 MO scheme of a complex with ligands without π contribution *(middle)*, with π donor ligands (π bases, *left*) and with π acceptor ligands (π acids, *right*). The π bonding behavior has a clear influence on the splitting Δ_O of the ligand field. π acceptor ligands lead to an increase of Δ_O, which is why, for example, the carbonyl ligand belongs to the strong-field ligands in the spectrochemical series

5. What effects do you expect for complex formation if you use 4d or 5d metals instead of 3d metals? What effects do you expect for complex formation if you increase the formal oxidation number of the central metal?
6. Are there π donor bonds or π acceptor bonds between the metal centre and the ligands in the following compounds? [Pt(C≡CPh)Cl(PPh$_3$)$_2$], [WO(OC$_2$H$_5$)$_4$], [Mo(NMe$_2$)$_5$], [Re(CH$_3$)(CO)]$_5$
7. The M-CO bond in transition metals in low oxidation states is often characterized by high stability. Discuss the bond type of the M-CO bond and explain this effect!

Color of Coordination Compounds

5

In Werner's day, the fascination with complexes certainly had a lot to do with their color. At those times the complexes that were already known (even if not recognized as such), such as the Berlin Blue, were captivating mainly (as the name suggests) because of their color. In this chapter we will learn where the color of complexes comes from and what we can learn about complexes from their color.

1. Where do the colors come from in coordination compounds?
2. Why is the intensity of the coloration sometimes very different?
3. What information can be gained from the d-d transitions?

5.1 Why Are Complexes Coloured?

Chromaticity occurs when a compound absorbs light of a certain wavelength range. The energy of the light causes an electron to leave an occupied energy level and occupy a previously empty (or half-occupied) level. The human eye sees the complementary color associated with that wavelength. Figure 5.1 shows the optical spectrum of light with the corresponding wavelengths and complementary colors.

The color variety of transition metal ions and their compounds forms a contrast to the usually colorless ions of the main groups. This leads to the conclusion that there must be a connection between color and the electrons in the d-orbitals. The ligand field theory provides us with an explanation for the colorfulness of many coordination compounds. By energetically splitting the d orbitals in the ligand field, an electron can be lifted from an occupied lower d orbital to an energetically higher d orbital. To illustrate this, we consider an octahedral d^1 complex – the pale violet hexaaquatitan(III) is formed upon reduction of

B. Weber, *Coordination Chemistry*,
https://doi.org/10.1007/978-3-662-66441-4_5

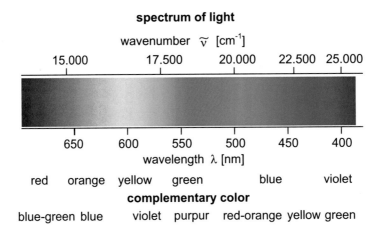

Fig. 5.1 Colours of the optical spectrum with wavelength range [nm] as well as complementary color after absorption of the color in the corresponding wavelength range

titanium(IV) in aqueous medium (to a suspension of $TiOSO_4$ in dil. H_2SO_4 zinc powder is added, Fig. 5.2).

The color of the titanium(III) ion is due to a so-called d-d transition of a valence electron from the energetically lower t_{2g} level to the higher e_g level. Because of the context $h \cdot \nu = \Delta E$, the color of the compound depends on the energy difference between the d orbitals, i.e., on the ligand field splitting Δ_O. Thus, in this simplest case, Δ_O alone determines the coloration of the compound. As we learned in Sect. 4.3, this is strongly determined by the position of the central ion and the ligand in the spectrochemical series as well as by the number of ligands and the coordination geometry. This is illustrated by the next examples. The first example is the color of chromium(III) or chromium(II) ions in water, Fig. 5.3 on the left shows the solutions of the two complex cations $[Cr(H_2O)_6]^{3+}$ and $[Cr(H_2O)_6]^{2+}$. Both complexes are octahedral and have the same ligand – water. The difference is the oxidation state of the central ion, it is a chromium(II) complex and a chromium(III) complex, respectively. The different position of chromium in the oxidation states +2 and +3 in the spectrochemical series leads to different values Δ_O for the splitting of the d orbitals in the ligand field and thus to a different color. In the next example, the metal center remains the same. An aqueous solution of $Co(NO_3)_2$ is pink, but the solution in concentrated hydrochloric acid is blue (Fig. 5.3 middle). According to the spectrochemical series, both water and the chloride ion are weak-field ligands and such a pronounced color difference would not be expected here. The answer to this observation is the different coordination number in the two complexes. In water, the octahedral complex $[Co(H_2O)_6]^{2+}$ is formed, while in concentrated hydrochloric acid, the tetrahedral anion $[CoCl_4]^{2-}$ is formed. In Chap. 4 on the ligand field theory, we have already discussed that due to the smaller number of ligands (four instead of six), the splitting in the tetrahedral ligand field is equivalent to $\Delta_T = \frac{4}{9}\Delta_O$ for the same central ion and the same (or in this case comparable)

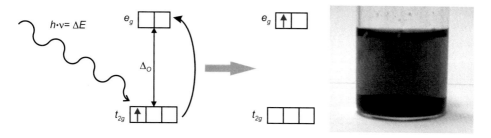

Fig. 5.2 Electronic transition between d-orbitals and color of the violet Ti^{3+}-ion

Fig. 5.3 *Left:* Color of $[Cr(H_2O)_6]^{3+}$ *(left)* and $[Cr(H_2O)_6]^{2+}$ *(right)*. *Middle:* Color of $Co(NO_3)_2$ in water *(left)* and in hydrochloric acid *(right)*. *Right:* color of $[FeCl_n(H_2O)_{6-n}]^{3-n}$, $[Fe(H_2O)_{6-n}(SCN)_n]^{3-n}$ with n = 1–3 and $[FeF_6]^{3-}$

ligands. The much smaller splitting in the case of tetrachloridocobaltate(II) leads to the absorption of the longer wavelength (lower energy) orange light and we see the blue complementary color. In the case of hexaaquacobalt(II), shorter wavelength (higher energy) green light is absorbed and we see the complementary color red.

The different splitting of the ligand field provides an explanation for the many different colors we observe in complexes. The next question we ask is why $[FeF_6]^{3-}$ and $[Fe(H_2O)_6]^{3+}$ are colorless. The three different iron(III) complexes $[FeCl_n(H_2O)_{6-n}]^{3-n}$, $[Fe(H_2O)_{6-n}(SCN)]_n^{3-n}$ with n = 1–3 and $[FeF_6]^{3-}$ are given in Fig. 5.3 on the right. All ligands have a similar position in the spectrochemical series (weak-field ligands) and lead to the formation of a high-spin complex. To find an answer to this, we need to look at the selection rules for electronic transitions. These selection rules also explain why the intensity of the color is different for different complexes.

Table 5.1 Molar extinction coefficient as a function of the electronic transition [3]

Electronic transition	ε [$l\ mol^{-1}\ cm^{-1}$]
d-d transition O_h	1–10
d-d transition T_d	10^2–10^3
CT transition	10^3–10^6

5.2   Selection Rules for Electronic Transitions

In addition to the d-d transitions just introduced, two other transitions are possible for complexes. The charge-transfer (CT) transitions and ligand-ligand transitions. In the latter, the metal centre plays a minor role and the color of the ligand is considered. It is often due to π-π^* transitions and will not be considered further. In the discussion of the selection rules in the remainder of this section, we focus on the electronic transitions where the metal center plays a major role. Experimentally, the intensity of the transition (i.e., the intensity of the color) can be used to estimate which of the transitions is observed. The molar extinction coefficient ε indicates how much light is absorbed by a 1-molar solution that is 1 cm thick. The larger ε, the more intense the color. Table 5.1 shows the magnitude of the molar extinction coefficient as a function of the electronic transition [3].

In order to be able to explain the differences in the various transitions, the spectroscopic selection rules have to be taken into account, which state the conditions whether an excitation by the available radiation is physically possible (i.e. allowed). Here, two rules are to be queried: the spin selection rule and the Laporte rule:

The **spin selection rule** states that the multiplicity M the system must not change during the electronic transition. Since $M = 2S + 1$, this depends on the total spin S. This does not change if the spin of the electron is preserved during the electronic transition (prohibition of spin reversal). The already mentioned example complexes $[FeF_6]^{3-}$ and $[Fe(H_2O)_6]^{3+}$ are both iron(III) complexes with weakfield ligands. Thus, the iron center is in the high-spin state and each of the five d electrons occupies one of the five d orbitals by itself, the occupation of the orbitals follows Hund's rule (S is maximal). Thus, all electrons have the same spin. In the case of a d-d transition, an electron would now have to be transferred from one of the lower d orbitals to one of the higher d orbitals. For this to happen, the spin would have to reverse due to the Pauli principle (no two electrons in an atom can have the same spin in all four quantum numbers). However, this is exactly what the spin selection rule prohibits us from doing, which is why d-d transitions are forbidden in such cases. The compound (or complex) appears colorless to us. In fact, d-d transitions can be found in highly concentrated solutions, but they are extremely weak ($\varepsilon \ll lmol^{-1}cm^{-1}$). Also, the spin selection rule obviously does not hold absolutely. For our iron(III) complexes given in Fig. 5.3, this means that no d-d transitions occur in all three examples. Nevertheless, only the hexafluoridoferrate(III) ion and the hexaaquairon(III) ion (not shown) are colorless. Thus, we can conclude that for the other two complexes, the coloration is caused by other transitions. It should be noted that the prohibition of spin reversal does not apply to all iron(III) complexes. In the case of the hexacyanidoferrate(III) – which is also an iron(III) complex – we have a strong-field ligand with the cyanide ion and the complex is in the low-spin state. Here, a d-d transition can take place without touching the spin selection rule.

The **Laporte rule** can be conveniently split into two queries:

1. Does the parity of the orbital change during the electronic transition? The parity describes the behavior of the orbital during point reflection. So we have to consider whether the orbital has an inversion center. If it remains unchanged during point reflection and consequently has a center of inversion, the parity is "gerade" (german for even), or g for short. As we can easily see in Fig. 5.4, this is the case for s and d orbitals. If the orbital lobes change s signs during point reflection and therefore have no inversion center, the parity is "ungerade" (german for odd), or u for short. This is the case for p and f orbitals. Excitation by light is only Laporte allowed if the parity changes, e.g. if an electron is excited from a p to a d orbital. Crystal field transitions in which electrons are excited from a d to a d orbital are in principle forbidden for the time being. However, as we have seen from the examples discussed so far, this prohibition cannot be as strict as the prohibition of spin reversal – otherwise we would hardly have colored complexes! This brings us to the second query of the Laporte rule:

2. Is the complex centrosymmetric (inversion symmetric; symmetric with respect to the point reflection)? If no, for example in tetrahedral and other non-centrosymmetric complexes, then excitation is allowed. If yes, for example in octahedral and other centrosymmetric complexes, then excitation is forbidden. That octahedral complexes, such as the $[Co^{II}(H_2O)_6]^{2+}$-ion, show an color at all, is due to processes that touch the Laporte rule. Among the vibrations that the ligands perform relative to the central metal are those that briefly cancel centrosymmetry, making electron excitation possible but then much weaker. Covalent bond fractions in complexes can also lead to weakening of the Laporte ban.

In summary, if an electronic transition is spin forbidden, then it will only occur with extremely low probability. A classic example of spin forbidden electronic transitions are iron(III) high-spin complexes. If a transition is Laporte forbidden, then this leads to a

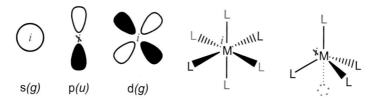

Fig. 5.4 Inversion symmetry for octahedral and tetrahedral complexes and for s, p, and d orbitals with the resulting parity, denoted as g from *gerade* (german for even) and u from *ungerade* (german for odd)

weakening of the intensity of the color. A transition that is allowed in the query for the change in parity of the orbital is intense and has a large extinction coefficient. This is the case, for example, with charge-transfer transitions, which often occur in complexes by excitation of electrons from occupied p orbitals into empty or partially occupied d orbitals. These charge-transfer transitions are responsible, for example, for the coloration of the iron(III) complexes $[FeCl(H_2O)_5]^{2+}$ and $[Fe(H_2O)_3(SCN)_3]$ given in Fig. 5.3. In both, the iron is present in the high-spin state. Parity forbidden transitions generally have lower intensity. If the complex is then also centrosymmetric, which is the case for octahedral complexes, the transition is particularly weak. This is visible for the cobalt complexes shown in Fig. 5.3. The blue of the tetrahedral tetrachloridocobaltate is more intense than the pink of the octahedral aqua complex. The fact that electronic transitions occur in an octahedral complex at all is due to the fact that a complex is not a rigid structure, but the metal-ligand bonds oscillate around an equilibrium distance. This momentarily removes the centrosymmetry and in this time window the electronic transition is possible. The transition occurs, but not as frequently as in tetrahedral complexes and is therefore weaker.

5.3   Charge Transfer (CT) Transitions

For a charge-transfer (CT) transition, an electron is excited from an (odd) p orbital into an (even) d orbital, so the transition is Laporte-allowed also for a centrosymmetric complex. In addition to ligand→ metal CT transitions (LMCT), there are also transitions of electrons from occupied d orbitals into empty ligand orbitals (metal→ ligand CT transition, MLCT). Since CT transitions can always be formulated in spin-allowed terms, CT absorptions are intense. They are particularly noticeable when d-d transitions are spin- and Laporte-forbidden, i.e., the compound does not exhibit significant crystal field color. Examples are high-spin iron(III) compounds such as $FeCl_3 \cdot 6H_2O$ (orange-yellow), $FeBr_3$ (dark brown) or the deep red thiocyanato complex $[Fe(H_2O)_3(SCN)_3]$ used for iron(III) detection. Furthermore, the intense CT transitions are noticeable where crystal field transitions are not possible due to the absence of d electrons. An example is the deep purple color of the d^0 ion permanganate, MnO_4^-. Now the question arises why CT transitions are observed for some compounds and not for others. An entry point to this question is provided by the iron(III) salts already mentioned. The corresponding sulfate or fluoride is colorless, the chloride is yellow-orange, the bromide is dark brown, and the corresponding iodide is not stable (Fig. 5.5). This is because the iodine has a lower electron affinity than the other halides because of its large radius, accordingly the ionization energy is low. The iodide ion is therefore more willing to donate its electron in a redox reaction. The ferric iodide decays to ferrous iodide and elemental iodine. The observed trend for the intensity of the CT transition mirrors the course of electron affinity for the halogens in the periodic table. In the Fluoride < chloride < bromide < iodide series, it becomes increasingly easier to

Fig. 5.5 *Left:* Ferric sulphate, ferric chloride and ferric bromide. *Right:* aqueous solution of vanadate, chromate and permanganate ion

oxidize the halide. The excited state produced by the CT excitation has a lower and lower energy until the electron transfer occurs directly. The excited state becomes the ground state and electron donor and acceptor react already in the absence of light in a redox reaction under electron transfer – thus *charge transfer*. Although we are dealing with solids here, the comparison is valid. In the case of iron(III) chloride and iron(III) bromide, the iron has the coordination number 6 and is octahedrally surrounded by six halide ions – just as in the case of the hexafluoride ferrate(III) ion.

The same concept can be applied to the colours of vanadate (VO_4^{3-}), chromate (CrO_4^{2-}) and permanganate (MnO_4^-). In the case of the three ions, the metal is present in its highest possible oxidation state corresponding to the group number. Thus, it has no d electrons which could be responsible for a d-d transition. Accordingly, the vanadate ion is colorless. However, an aqueous solution of chromate or permanganate ions is yellow or deep purple, respectively, as shown in Fig. 5.5. The reason for this can only be CT transitions from occupied p orbitals of the O^{2-} ion to the empty d orbitals of the metal center. In this case, the intensity of the color correlates with the oxidation power of the metal center. Manganese in the +7 oxidation state is a much stronger oxidant than chromium in the +6 oxidation state, while for vanadium +5 is the most stable oxidation state. The tendency of the metal center to accept electrons correlates with the intensity of the transition and the energetic position of the CT band. Thus, for the yellow color, short-wavelength high-energy blue light is absorbed by the chromate ion, and for the purple color, much longer-wavelength (and thus lower-energy) green light is absorbed by the permanganate ion. Moving from permanganate to the higher homologues in the periodic table (the corresponding 4d and 5d elements), we know from lectures or textbooks on general chemistry that the transition to the higher homologues increases the stability of the maximum possible oxidation state (because the outer electrons are shielded from the positively charged nucleus by more electron shells, making them easier to remove). This trend is reflected in the coloration of the corresponding ions. Thus, the (radioactive) TcO_4^- is yellowish, while the perrhenate ReO_4^- is colorless. For the central ions of the 4d and 5d series, more energy is required for the CT transition and the CT bands move successively into the UV region.

5.4   d-d Transitions and the Determination of Δ_O

For d-d transitions, a d electron of the central ion changes from a lower to a higher energy level according to the selection rules. So far, we have assumed that the energy of the incident light corresponds to the energy difference between the d orbitals. For octahedral complexes, this would be the ligand field splitting Δ_O, which thus determines which color the complexes have. The other way around, one could then conclude that it is possible to determine Δ_O from the color of the complex. A look at the different colors of divalent 3d elements in aqueous solution shows that this is not quite so simple. The copper(II), nickel (II) and cobalt(II) ions are present in water as hexaaqua complexes and the observed colours for the ions are blue, green and pink (see Fig. 5.6). The divalent ions are close together in the spectrochemical series, have the same ligands and the same coordination environment. Thus, one would expect Δ_O to be in a similar range for the three complexes, which is a contradiction to the different colors. When looking at the UV-Vis spectra of the solutions one finds that a different number of absorption bands are present. The observed color is the sum of the individual bands.

Indeed, a determination of Δ_O is possible with the help of the UV-Vis spectra, but not as simple as it seems at first sight. The assumption just made that the energy difference ΔE of the electronic transition corresponds directly to the ligand field splitting Δ_O is only realized in the so-called strong field approach (strong field approximation). Here we assume that the electrostatic interaction between the electrons in the d orbitals is negligible. This does not coincide with reality and observed optical spectra. The strong field is a hypothetical reference point.

As an example for the interpretation of the spectra and the determination of Δ_O we consider a spin-crossover compound. This is an octahedral iron(II) complex which is in the low-spin (LS) state at low temperatures and in the high-spin (HS) state at higher temperatures. All further details on this phenomenon are given in the magnetism section. It is important for the understanding of the coloration that the metal-ligand bond length increases during the transition from the LS to the HS state, because in the HS state the antibonding e_g orbitals are occupied. This leads according to the formula

Fig. 5.6 Aqueous solution of a copper(II), cobalt(II) and nickel(II) salt. The three ions are each present as a hexaaqua complex

$$10Dq(r) = 10Dq(r_0)\left(\frac{r_0}{r}\right)^6$$

to a change in the ligand field splitting Δ_O and thus to a color change during the spin transition. In addition to the parameter Δ_O, which balances the interactions of the metal electrons with the ligand electrons in a complex, we need a second parameter for the following discussion, the Racah parameter B. This balances the interactions of the metal electrons with each other. So we leave the strong field approach and go to the so-called weak field approach (approximation of the weak field), where the electrostatic interactions between the d electrons play the dominant role. B is a defined value for the free atom or ion (B_0), but in the presence of ligands it becomes the second parameter of the ligand field theory, since B depends on the type and number of ligands, where always: $B < B_0$. Complexation reduces the interaction of the metal electrons with each other, we are in a smooth transition between the weak field and the strong field approach. Please note that you cannot equate the strong-field and weak-field approach with strong-field and weak-field ligands.

The dependence of the chromaticity on the metal-ligand distance and thus on Δ_O is not only seen in spin crossover. Similar effects can be observed in solids. An example is the color of the halides of cobalt. The (anhydrous) salts CoF_2, $CoCl_2$ and $CoBr_2$ have the colors pink, blue and green. The different colors cannot be explained by charge-transfer transitions in the visible, in contrast to the halides of the iron(III) ion. For this, the metal center does not have the correct oxidation state. Apart from the fluoride, the corresponding cobalt(III) salts do not exist as stable compounds for the same reasons as for the iron(III) iodide. The series CuF_2, $CuCl_2$, $CuBr_2$ and CuI_2 also show the phenomena discussed for the halides of the ferric ion, up to the decomposition of the iodide to iodine and copper (I) iodide. The cobalt(II) ion has the same coordination environment in all salts, the coordination number is 6, it is located in the octahedral gaps of the densely or most densely packed anions. If we now disregard the different packing patterns, the only parameter that changes continuously is the size of the anion. We can now assume that as the ion radius increases, the octahedral gap increase in size. Since our cobalt(II) ion always has the same size, this leads to an increase in the cobalt-halide distance, which we equated with the metal-ligand distance. Consequently, for cobalt halides, as the ionic radius for the anion increases, the splitting of the d orbitals for the cobalt (octahedral ligand field) should decrease. If we look at the complementary colors of the halides – green for the fluoride (we see pink), orange for the chloride (we see blue), and red for the bromide (we see green), then indeed the wavelength of the absorbed radiation increases. The absorbed light has less and less energy, and the splitting of the d orbitals becomes smaller and smaller. The same effect is responsible for the red color of ruby – a chromium(III) ion doped (1–8%) α-Al_2O_3 structure. The ionic radius of the chromium(III) ion, 0.755 Å, is larger than that of the aluminium(III) ion, 0.675 Å. Chromium(III) compounds are often green, a good example being Cr_2O_3, which also has the α-Al_2O_3 structure. In green chromium(III) oxide, light is absorbed in the red wavelength range. In ruby, due to the smaller ionic radius from the

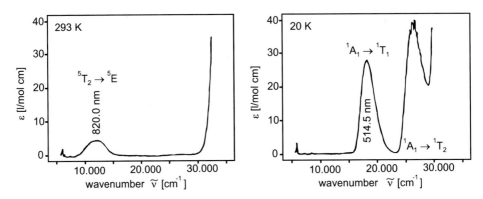

Fig. 5.7 Single crystal absorption spectra of the spin crossover complex [Fe(ptz)$_6$](BF$_4$)$_2$ at 295 K *(left)* and 20 K *(right)*. (Adapted from Ref. [19])

aluminum(III) ion, the metal-ligand distance for the chromium(III) ion is shorter and, accordingly, the energetic splitting of the d orbitals is larger. Here, the light is absorbed in the shorter-wavelength green region and we see the red complementary color associated with it.

Returning to the iron(II) complexes, we consider the absorption spectra of the spin-crossover complex [Fe(ptz)$_6$](BF$_4$)$_2$ at 295 K (HS state) and 20 K (LS state), given in Fig. 5.7 [19]. In the HS state, the complex is colorless. In the UV-Vis spectrum, a band is observed at 820 nm (12,195 cm^{-1}). The transition is in principle similar to the d-d transition of the titanium(III) ion, with the difference that five more d electrons are present. Since the HS state is present, the five electrons are evenly distributed among the five d orbitals and the sixth electron is in the ground state in one of the three t_{2g} orbitals. Light absorption leads to an excited state in which one of the energetically raised e_g orbitals is occupied. In the LS state, the complex is red due to the color change discussed earlier. In the absorption spectrum, an associated band is observed at 514.5 nm (19,436 cm^{-1}). However, a second band of similar intensity at higher energy (shorter wavelength or larger wavenumber) is seen right next to it. The reason for this is the already announced interaction of the metal electrons with each other, the Racah parameter B. To understand this, we need to take a closer look at the electron configuration of the excited state. In the case of the iron(II) complex in the LS state, the six electrons are in the three t_{2g} orbitals, each of which is doubly occupied. Now, if an electron is excited from, say, the d_{xy} orbital, it can be accommodated in the $d_{x^2-y^2}$ orbital or the d_{z^2} orbital in the excited state. At first glance, it seems to make no difference. However, this is not the case. The $d_{x^2-y^2}$-orbital is located in the x-y plane, just like the d_{xy} orbital, while the preferred direction of the d_{z^2} orbital is along the z axis. If this is occupied, there is an increased repulsion between the electrons in the d orbitals with z fraction – the excited state formed in this way is energetically higher than the variant in which the $d_{x^2-y^2}$ orbital is occupied and the repulsion between the electrons of the d orbitals does not change significantly. Nevertheless, to determine the ligand field splitting energy from the spectra, we need the Tanabe-Sugano diagrams.

5.4.1   Tanabe-Sugano Diagrams

The examples given so far have shown that the colour of complexes based on d-d transitions depends on the energy difference between the d-orbitals, i. e. on the ligand field splitting Δ_O, according to the relation $h \cdot \nu = \Delta E$. From the UV-Vis spectra of complexes, ν and hence ΔE can be determined. However, this is not as simple as it appears at first sight, because in complexes with more than one d electron different electronic transitions take place, leading to several bands with different ν. To make the assignment possible, *Tanabe* and *Sugano* calculated the dependence of the term energies of octahedral complexes on the ligand field strength in the 1950s. They presented their results in diagrams in which the ligand field strength and term energy are given in units of the effective Racah parameter B'. In these Tanabe-Sugano diagrams with their plot of E/B' versus Δ/B', the term energies themselves are not given, but the differences from the ground state energy. Thus, the energy of the ground state always corresponds to the x axis!

   In order to understand the diagrams, we must first deal with the term concept already introduced in Chap. 4.

   Effect of a ligand field on unperturbed terms The electron configuration often only incompletely describes the electronic state of an ion. This is because a given configuration can lead to several atomic microstates that are of the same or different energy, depending on whether the electron-electron repulsion is the same or different. For example, for a given configuration, orbitals with different orbital angular momentum orientations (characterized by a different orbital angular momentum quantum number m_l) may be occupied. In addition, different spin orientations (characterized by the spin quantum number m_s) are also possible. A group of microstates with the same values of total orbital angular momentum L and total spin S belonging to a given configuration is called a term. To a given total orbital angular momentum quantum number L belong $2L + 1$ states taking values between $-L$ and L. In the Tanabe-Sugano diagram in Fig. 5.8, on the far left at $10\ Dq = 0$, the field-free state of a d^6 ion is given. The ground state has the term symbol ^5D. D stands for the total orbital angular momentum $L = 2$, which implies that the state is five times orbitally degenerate. The superscript 5 stands for the spin multiplicity $2S + 1$, meaning the total spin of the system is 2 (or $\frac{4}{2}$ = four unpaired electrons). The precise rules for determining the term symbols unperturbed from the ligand field have already been presented in Chap. 4. In fact, there are five ways to arrange the six d electrons among the five d orbitals under these specifications, as given in Fig. 5.9 on the left.

   If the ion is placed in a ligand field, the terms determined for the absence of a ligand field split energetically up, and the energy of the in this way obtained terms (or molecular symmetry labels) is given as a function of Δ in the Tanabe-Sugano diagrams. Staying with the iron(II) ion in the octahedral ligand field, we find that for the HS state there are once three and once two microstates of equal energy. These can be grouped into two terms. Here the term symbol indicates the degree of degeneracy. Terms of type T are threefold degenerate, A denotes a nondegenerate state and E a twofold degenerate state. The multiplicity is again given in the upper left, which does not change. Due to the splitting of the d orbitals in the ligand field, the three microstates in which the sixth electron is

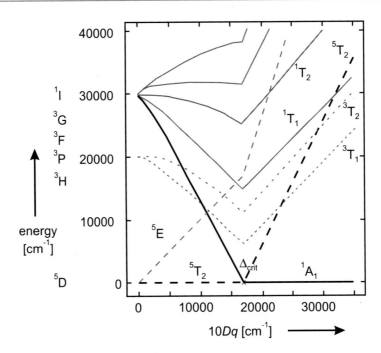

Fig. 5.8 Tanabe-Sugano diagram of a d^6 system in the octahedral ligand field. For clarity, E and $10\,Dq$ (Δ) have been used for the axis labels. For iron(II), the electron interaction parameter is $B' \approx 1050$ cm^{-1}.

accommodated in one of the energetically more favorable t_{2g} orbitals belong to the ground state $^5T_2$. The other two microstates in which the sixth electron is accommodated in one of the two energetically higher e_g orbitals belong to the excited state ^5E.

When a critical ligand field strength Δ_{crit} is exceeded, the ground state is no longer the $^5T_2$ state, but the $^1A_1$ state. This term belongs to the field-free term ^1I. I corresponds to a total orbital angular momentum of $L = 6$, which implies that this state is 13-fold degenerate. Here the attentive reader is required to write down the 13 microstates. For further discussion, we only need the seven given in Fig. 5.9 on the right, which are composed of the $^1A_1$ ground state and the excited states $^1T_1$ and $^1T_2$.

In all three cases the total spin is $S = 0$. In the field-free state, these terms are energetically higher than the ^5D term, since the spin pairing energy must be applied here. As mentioned earlier, $^1A_1$ is the ground state in the case of a high ligand field splitting, and the other two states are excited states, each of them has an electron lifted from a t_{2g} orbital into an e_g orbital. $^1T_2$ is energetically higher than $^1T_1$, because of the additional repulsion between electrons of spatially close d orbitals discussed earlier (e.g. xy and $x^2 - y^2$). For this reason, the Tanabe-Sugano diagrams take both parameters, the ligand field splitting Δ_O and the Racah parameter B, into account.

In the framework of crystal field theory and ligand field theory, we start in the weak field approximation; thus, we assume that the interactions of the metal electrons with each other

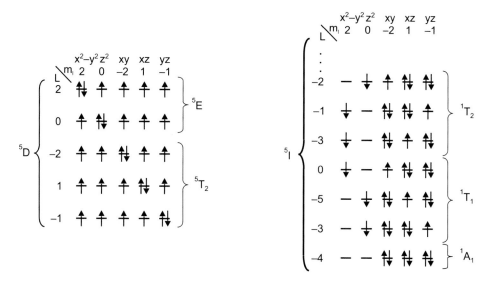

Fig. 5.9 Microstates of the iron(II) ion in the high-spin *(left)* and low-spin *(right)* states with associated terms. For the low-spin state, only the microstates and terms relevant for the following discussion are given

are fundamentally stronger than those with the ligand electrons. In this picture, starting from the spectroscopic terms of the field-free ions, we can introduce the ligand-metal interactions as perturbations according to Δ_O. Thus, the presented model proves to be very useful in the interpretation of the properties of complexes and their electron spectra.

With the help of the Tanabe-Sugano diagrams, the bands of the electron spectra can be assigned to individual transitions and the ligand field splitting Δ can be determined. Complete Tanabe-Sugano diagrams contain all conceivable excitations, including spin forbidden ones. To be able to interpret the stronger absorptions, it is sufficient to consider only the spin-allowed transitions, i.e., those in which the total spin does not change with respect to the ground state. With this background knowledge, we return to the UV-Vis spectrum of $[Fe(ptz)_6](BF_4)_2$ (ptz = 1-propyltetrazole) shown in Fig. 5.7. The bands at 820 nm and 514.5 nm (12,195 cm^{-1} and 19,436 cm^{-1}) can be assigned to the $^5 T_2 \rightarrow {}^5 E$ and $^1 A_1 \rightarrow {}^1 T_1$ transitions, respectively, in the Tanabe-Sugano diagram. If one enters the transitions, one obtains as 10 *Dq*-value for the HS-state 12,250 cm^{-1} and for the LS-state 20,050 cm^{-1}. It can be seen very nicely how the ligand field strength and thus the color of the compound changes during the spin transition due to the different bond lengths.

5.4.2 *trans*-[CoCl₂(en)₂]Cl and *cis*-[CoCl₂(en)₂]Cl

As a final example, we consider the two cobalt(III) complexes *trans*- and *cis*-[CoCl₂(en)₂] Cl. In Fig. 5.10, the UV-Vis spectrum of the two complexes and an image of the two

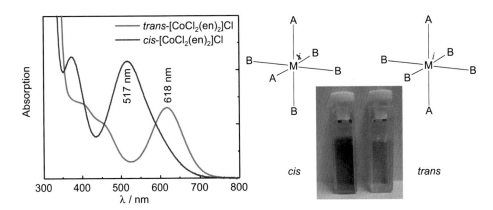

Fig. 5.10 UV-Vis spectrum and photograph of an aqueous solution of the green complex *trans-*[CoCl$_2$(en)$_2$]Cl. When the solution is heated to 70 °C for 5 min, the red complex *cis-*[CoCl$_2$(en)$_2$]Cl is formed, which is then stable even after cooling at room temperature

solutions are given. The cobalt(III) ion is also a 3d^6 system and both isomers are present in the low-spin state. The similarity to the iron(II) complex [Fe(ptz)$_6$](BF$_4$)$_2$ in the low-spin state is well seen, especially for the *cis* form (see Figs. 5.10 and 5.7). The colour of the two complexes is very similar and the UV-Vis spectra are also very similar in terms of the number and position of the bands. In comparison, the bands in the green *trans-*[CoCl$_2$(en)$_2$] Cl are significantly shifted towards longer wavelengths. Apparently, the two *trans-*chlorido ligands cause a slightly smaller splitting Δ_O. Another striking difference is the intensity of the bands. The *cis* complex has a larger absorption coefficient than the corresponding *trans* complex. This difference can be explained by the symmetry of the two isomers. During the transition from *trans* to *cis,* the inversion center at the metal center is removed. Although it is an octahedral complex, the transition is Laporte-allowed, at least according to the second query, and has as a result a higher extinction coefficient than the corresponding *trans* complex.

5.5   Questions

1. Why is [Fe(H$_2$O)$_6$]$^{3+}$ colorless and [Fe(H$_2$O)$_6$]$^{2+}$ pale green?
2. Why do the color and intensity of Co^{2+} differ in aqueous solution (pink, hexaaquacobalt (II)) and in concentrated hydrochloric acid (blue, tetrachloridocobaltate(II))?
3. To mask Fe^{3+} one uses fluoride ions. What happens?
4. An aqueous solution of Cu^{2+} is pale blue, on addition of NH$_3$ it becomes deep blue. Explain the observation!
5. Determine the term symbol and the possible ligand field split terms from the ground state of a d^2-system in the octahedral ligand field.

Stability of Coordination Compounds

6

In describing ligands and their complexes, the term stability has been mentioned several times, so in this chapter we ask the question:

What does "a complex is stable" mean?

6.1 What Is a Stable Complex?

The different stability of complexes is illustrated by the reaction of iron(III) ions with the ligands chloride, thiocyanate and fluride in water. For this purpose, a solution of ferric nitrate (the cations $[Fe(H_2O)_5(OH)]^{2+}$ and $[Fe(H_2O)_6]^{3+}$ are present, the solution is yellowish) is acidified with dilute nitric acid to give the colorless aqua complex $[Fe(H_2O)_6]^{3+}$. When chloride ions (conc. HCl) are added, the solution becomes yellow. Ligand exchange takes place and the cations $[FeCl_n(H_2O)_{6-n}]^{3-n}$ with $n = 1–3$ are formed (Fig. 6.1 left). Now some ammonium thiocyanate solution is added to part of the solution. The solution turns deep red. In the middle of Fig. 6.1 a highly diluted solution is shown, where the red color is clearly visible. When the thiocyanate ions are added, a ligand exchange takes place, replacing the chloride ions with thiocyanate to give the complexes $[Fe(H_2O)_{6-n}(SCN)_n]^{3-n}$ with $n = 1–3$. A spatula tip of sodium fluoride is now added to this solution and after stirring briefly the solution becomes colorless again (Fig. 6.1 right). The resulting hexafluoridoferrate(III) complex is very stable and is used in analysis to "mask" iron(III) ions.

In order to understand the color changes described, let us look at the individual reaction equations for the processes just described. Already the first equation raises a question that needs to be answered.

© The Author(s), under exclusive license to Springer-Verlag GmbH, DE, part of Springer Nature 2023
B. Weber, *Coordination Chemistry*,
https://doi.org/10.1007/978-3-662-66441-4_6

Fig. 6.1 Aqueous solution of
iron(III) ions in the presence of
chloride ions *(left)*, after addition
of thiocyanate ions *(middle)* and
after further addition of fluoride
ions *(right)*

$$\left[Fe(H_2O)_5(OH)\right]^{2+} + H^+ \rightleftharpoons \left[Fe(H_2O)_6\right]^{3+}$$

Why does dissolving iron(III) nitrate in water produce the complex $[Fe(H_2O)_5(OH)]^{2+}$ and
not the corresponding hexaaquairon(III) complex?

To answer this question, we first check the pH of the solution. It is pH = 2–3, which
means that the solution is acidic. The reaction equation shows that the hexaaquairon(III)
ion obviously behaves like a Brønsted acid, a proton donor. Indeed, the pK_S value of [Fe
$(H_2O)_6]^{3+}$ is 2.46 [1]. For comparison, the pK_S value of acetic acid (as an example of a
weak acid) is 4.75, that of nitric acid (strong acid) is −1.37 and that of pure water is
14 [1]. The hexaaquairon(III) ion is a moderately strong acid, roughly comparable to
phosphoric acid (H_3PO_4, pK_S = 2.16). The stronger a Brønsted acid is, the higher is its
tendency to donate a proton.

In the previous chapters, we have primarily discussed how the ligands affect the
properties of the central ion. Which ligand coordinates at the metal ion, the coordination
number and complex geometry all affect the splitting of the d orbitals and hence the color
and magnetic properties of the complex. This is not a one-way process. The coordination of
the ligand to the metal also changes the properties of the ligand, which is then reflected in a
change in reactivity. This is the basis for metal-center-mediated catalytic processes!
Coordination of water to the ferric ion, which is a strong Lewis acid, weakens the O-H
bond in water and facilitates the release of protons. The pK_S value from the water changes.
If the pH of the solution is increased further, more deprotonation takes place. Condensation
reactions follow with the formation of polynuclear complexes (the so-called
isopolycations). A simple example of a dinuclear species is the cation $[\{Fe(H_2O)_4\}_2$
$(\mu\text{-}OH)_2]^{4+}$. We practice IUPAC nomenclature again, it is the bis(μ-hydroxido)bis
(tetraaquairon(III)) ion.

The behavior described for the iron(III) ion in aqueous solution is observed for all highly
charged (trivalent) small cations. Another example is the aluminium (III) ion. Here the pK_S
value of the hexaaqua complex is 4.97 [1] – it is a slightly weaker acid. The difference in

acidity between the two hexaaqua ions is the basis for the Bayer process, mentioned in the introduction, for the production of high purity alumina for the large-scale preparation of aluminum. It causes $Al(OH)_3$ to be amphoteric and $Fe(OH)_3$ not. The pK_S value of the bound water does not depend exclusively on the Lewis acid strength. An element is a stronger Lewis acidic, the more positively charged and smaller (radius) it is. Accordingly, the more negatively charged and smaller an element is, the more Lewis basic it is. The positive charge of the two ions is the same, they are both trivalent. The ionic radius of aluminum is smaller than that of iron (for CN 6: 0.675 Å vs. 0.785 Å), aluminum is the stronger Lewis acid.

We return to the stability of complexes and consider the following equations. Here we assume that the amount of ligands used is the same.

$$\left[Fe(H_2O)_6\right]^{3+} + Cl^- \rightleftharpoons \left[FeCl(H_2O)_5\right]^{2+} + H_2O$$
$$\left[FeCl(H_2O)_5\right]^{2+} + SCN^- \rightleftharpoons \left[Fe(H_2O)_5(SCN)\right]^{2+} + Cl^-$$
$$\left[Fe(H_2O)_5(SCN)\right]^{2+} + 6\,F^- \rightleftharpoons \left[FeF_6\right]^{3-} + SCN^- + 5\,H_2O$$

The experiment has shown that for the iron(III) complexes the stability in water increases as a function of the ligand in the order $H_2O < Cl^- < SCN^- < F^-$. A numerical value for direct comparisons is provided by the equilibrium constant K. It is the quotient of the concentration of the products and the concentration of the reactants. The general formula (for an octahedral complex) is:

$$M^{n+} + 6\,L \rightleftharpoons \left[M(L)_6\right]^{n+}$$

$$K_B = \frac{c\left(\left[M(L)_6\right]^{n+}\right)}{c(M^{n+}) \cdot c^6(L)}$$

If the numerical value is large, then the equilibrium is on the side of the product, which in this case is the complex. If the numerical value is small (significantly less than 1), then the equilibrium is on the side of the starting materials. The concentration of the individual components plays a major role here. The ferric thiocyanate complex can be formed from the hexafluoridoferrate ion if the excess of thiocyanate ions relative to the fluoride ions is large enough. Assuming that we are working with water as the solvent, we must remember that the metal ion is present as an aqua complex and that the reaction involves ligand exchange of water for the ligand used. Additional water is released in the process. Since it is the solvent, it does not matter in determining the equilibrium constant. The amount of water changes only insignificantly during the reaction and is therefore not taken into account. We look at the reaction using the example of the formation of the complex $[Fe(CN)_6]^{4-}$

$$Fe^{2+} + 6\,CN^- \rightleftharpoons \left[Fe(CN)_6\right]^{4-}$$

$$K = \frac{c\left(\left[Fe(CN)_6\right]^{4-}\right)}{c\left(Fe^{2+}\right) \cdot c^6(CN^-)}$$

In order to determine the equilibrium constant, the concentrations of the individual species in solution must be determined at equilibrium. This can be done, for example, if concentration-dependent UV-Vis spectroscopy is carried out. If the cyanide ion concentration is slowly increased, the stability constants for the successive addition of the ligands can also be determined, the product of which is again the equilibrium constant. The reciprocal of the equilibrium constant for complex formation is the dissociation constant ($K_D = 1/K$). For the successive addition of ligands we obtain:

$$\left[Fe(H_2O)_6\right]^{2+} + CN^- \rightleftharpoons \left[Fe(CN)(H_2O)_5\right]^{1+} + H_2O$$

$$K_1 = \frac{c\left(\left[Fe(CN)(H_2O)_5\right]\right)}{c\left(\left[Fe(H_2O)_6\right]^{2+}\right) \cdot c(CN^-)}$$

$$\left[Fe(CN)(H_2O)_5\right]^{1+} + CN^- \rightleftharpoons \left[Fe(CN)_2(H_2O)_4\right] + H_2O$$

$$K_2 = \frac{c\left(\left[Fe(CN)_2(H_2O)_4\right]\right)}{c\left(\left[Fe(CN)(H_2O)_5\right]\right) \cdot c(CN^-)}$$

$$K = K_1 \cdot K_2 \cdot K_3 \cdot \ldots$$

Figure 6.2 gives an example of such a UV-Vis titration. It shows the spectra for the titration of an iron(III) complex with a macrocyclic tetradentate ligand and two monodentate ligands. The two monodentate ligands can be exchanged stepwise. On the right side the molecular structure of such complexes is given for illustration.

6.1.1   The HSAB Concept

The equilibrium constant for complex formation provides well comparable numerical values, but their determination is time-consuming. For a quick comparison of the complex stability, the *HSAB theory* (or *concept of Pearson*) developed by Pearson helps us. HSAB theory stands for *theory of hard and soft acids and bases*. The acids and bases in this concept are Lewis acids and bases. Thus, it can be applied very well to coordination compounds to estimate the stability of the bond between metal ion (Lewis acid) and ligand (Lewis base). Besides the stability of complexes, this principle can also be used to explain the occurrence of compounds in nature (oxidic or sulfidic). This is closely linked to the "stability" of the corresponding salts, i.e. their (as low as possible) solubility.

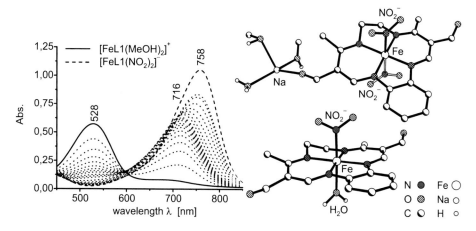

Fig. 6.2 UV-Vis spectra for the titration of [FeL1(MeOH)$_2$]$^+$ with nitrite. A successive exchange of methanol for nitrite occurs. This first gives the neutral complex [FeL1(MeOH)(NO$_2$)] and then the complex [FeL1(NO$_2$)$_2$]$^-$. L1 is a macrocyclic ligand. On the right side, the molecular structure of an analogous complex with two nitrite (top) and one nitrite and one water (bottom) as ligands is given [20]

The HSAB concept does not look at the strength or weakness of the acids and bases, but uses the concept of *hard* and *soft* to describe the stability of Lewis acid-base adducts. To do this, we must first explain what a hard and what a soft Lewis acid or Lewis base is.

Hard Lewis acids are small cations with a high positive charge. This means strong Lewis acids like Al^{3+} or Fe^{3+} are hard cations. However, ionic radius and charge are not the only factors that determine the hardness of a Lewis acid. Hard Lewis acids do not have non-bonding valence electrons, which means they are cations with a closed noble gas shell. A good example of this is the sodium ion and the copper(I) ion. Both ions have the same charge (+1) and a similar ionic radius (1.16 Å and 0.91 Å). The radius of the copper (I) ion is even smaller, it should be a harder Lewis acid. The valence electrons are decisive here. The sodium ion has a closed noble gas shell and is therefore hard. The copper(I) ion has a d^{10} shell and is soft. Soft Lewis acids are usually large cations with a small positive charge and valence electrons.

In the case of ligands, the Lewis bases, the hardness is predominantly determined by the donor atoms. Possible substituents play a subordinate role. It makes little difference whether water (H$_2$O), the hydroxide ion (OH$^-$) or the oxide ion (O^{2-}) act as ligand, the hardness of the three ligands is similar. The same trend holds for the donor atoms as for the cations. Hard Lewis bases have a small ionic radius. The charge plays a minor role. Instead, the electronegativity of the element can be used to estimate the hardness of the Lewis base.

The "hardness" is a measure of the polarizability of the ions or ligands. Hard ions are difficult to polarize (small, high oxidation state), while soft ions are easy to polarize. Table 6.1 gives an overview of the assignment of some ions and donor atoms.

Table 6.1 Relative hardness of some ions and donor atoms for the estimation of complex stabilities according to the HSAB concept

	Hard	Medium	Soft
Cations	H^+		
	Li^+, Be^{2+}, B^{3+}, C^{4+},	Fe^{2+}	Ni^{2+}, Cu^+, Zn^{2+}, Ga^{3+}, Ge^{2+}
	Na^+, Mg^{2+}, Al^{3+}, Si^{4+},	Mn^{2+}	Pd^{2+}, Ag^+, Cd^{2+}, In^{3+}, Sn^{2+}
	K^+, Ca^{2+}, Sc^{3+}, Ti^{4+},		Pt^{2+}, Au^+, Hg^{2+}, Tl^{3+}, Pb^{2+}
	(no d electrons)	(few d-electrons)	d^8/d^{10}
Donor atoms	F, O	N, Cl	Br, H^-, S, C, I, Se, P

The underlying idea of the HSAB concept is "like goes with like", i.e. hard ions form stable complexes with hard ligands and soft Lewis acids with soft Lewis bases. For the combination hard + hard, electrostatic interactions dominate the chemical bond. It has a pronounced ionic character. A good example of this is the poorly soluble CaF_2. For the combination soft + soft, the chemical bond has more covalent character, such as in the sparingly soluble HgI_2. The low solubility of salts represents strong attractive interactions between the anions and cations, that is, the Lewis bases and Lewis acids. The same trends as for complex formation are observed. Preferred coordination numbers and coordination polyhedra of complexes can also be explained. The hard sodium ion prefers the widely used coordination number 6 with octahedral coordination environment. The soft copper (I) ion preferentially forms bonds with covalent bonding moieties. For this reason, the preferred coordination number is 4 with a tetrahedron as the coordination polyhedron. This can be explained by the fact that only four orbitals are available to form a bond – one empty s and the three empty p orbitals.

Our experiment with iron(III) ions confirms the listed trends: The hard iron(III) ion does not form stable complexes with the relatively soft chloride ligands, so that the latter can easily be displaced by the somewhat harder thiocyanate (*N-donor*). However, the complexes are even more stable with the very hard fluoride ion.

A good example of the HSAB concept that occurs in the qualitative analysis practical course are the detection reactions for the halide ions. Here the different solubility of the silver halides can be explained using the HSAB concept. The solubility product decreases from silver fluoride (highly soluble) over silver chloride and silver bromide to silver iodide (increasingly unsoluble), reflecting the increasing softness of the halide anions (the Ag (I) ion is very soft!). This is noticeable when one tries to dissolve the precipitates, which is possible with diluted ammonia in the case of the chloride and concentrated ammonia in the case of the bromide (under complexation, see Chap. 1). In the case of iodide, a stronger complexing agent, thiosulfate (which is also known as a fixing salt in black-and-white photography), must be used.

6.2   Thermodynamic Stability and Inertness of Complexes

When we talk about the stability of complexes, we usually refer to their resistance to ligand substitution. Other alternatives, e.g. spontaneous decomposition into the elements, are not considered. Ligand exchange reactions are usually equilibrium reactions and an important question that arises when considering such reactions is on which side the equilibrium is, or in other words, whether the reaction proceeds or not. So far, we have described this question mathematically with the equilibrium constant and roughly compared the stability of the complexes with the HSAB concept. To answer this question in detail, we need to distinguish between the thermodynamic stability and the inertness or kinetic stability of complexes. The question of thermodynamic stability provides information on whether a reaction proceeds in the desired direction. To answer this question, we can determine the equilibrium constant or refer to the HSAB concept. The question of inertness tells us something about the activation enthalpy that must be overcome for the reaction to proceed at all. In the examples described so far, this aspect has been neglected. Figure 6.3 shows the difference. For the reaction between A and B (or for the reverse reaction between C and D) to occur at all, an energy barrier must be overcome, the activation energy or activation enthalpy E_A or ΔG^*. The hight of the energy barrier affects the reaction rate, which is temperature dependent. This relationship is given in the Arrhenius equation.

$$k = A \ \times \ e^{\frac{-E_A}{RT}}$$

In this equation, k is the rate constant, A is a pre-exponential factor, E_A is the activation energy, R is the gas constant, and T is the temperature in Kelvin. A very high energy barrier will result in the reaction not occurring. The following examples will show that kinetic stability also plays a role for ligand exchange reactions.

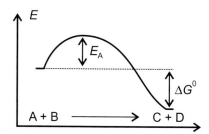

Fig. 6.3 Reaction diagram for the reaction of A + B to form C and D. For the reaction to occur, the activation energy E_A (often referred to as ΔG^*) must be overcome. The amount of this energy barrier affects the rate of reaction. Whether the equilibrium is on the side of the products (C and D) or the reactants (A and B) depends on the free energy ΔG^0. If the value is negative, then the reaction is exergonic and the equilibrium is on the side of the products. If the value is positive, then the reaction is endergonic and the equilibrium is on the side of the reactants. The frequently used term endothermic describes a reaction that only takes place with the addition of energy

A complex is thermodynamically stable when the difference in free energies ΔG^0 of reactants and products is negative (exergonic reaction). This difference is related to the extent of formation of a complex when equilibrium has been reached. Note the difference between the terms exergonic and exothermic. In an exergonic reaction, the equilibrium is on the side of the products and we consider ΔG^0. The term exothermic refers to the enthalpy of reaction ΔH. An exothermic reaction proceeds by releasing energy, while an endothermic reaction proceeds by absorbing energy.

Kinetic stability describes the speed with which a reaction moves towards the equilibrium setting. In the case of substitution reactions, it is always necessary to distinguish whether one is speaking of thermodynamic stability or kinetic stability. Furthermore, it is important to specify which reaction or reaction conditions the statement refers to. The term "a stable complex" can refer to a thermodynamically stable complex that is kinetically stable or labile, but it can also refer to a thermodynamically unstable complex that is kinetically very stable and therefore simply does not react. We look at the following examples to better understand the difference.

The first examples come from main group chemistry. The molecular compounds CCl_4, $SiCl_4$, and NF_3 are thermodynamically stable with respect to spontaneous decomposition into the elements, while NCl_3 spontaneously decomposes into its elements. But all the four examples are thermodynamically unstable with respect to hydrolysis. Silicon tetrachloride is a liquid that can be distilled without decomposition. However, when handling this compound, it must be done in the absence of moisture, otherwise immediately, due to the high affinity of silicon for oxygen, the compound will hydrolyze with HCl release. In contrast, carbon tetrachloride is much more stable and can be used as a solvent without concern (in terms of undesirable decomposition). This is surprising at first sight, since the C-Cl bond is weaker than the Si-Cl bond. The different kinetic stability of the two molecules can be explained by the Lewis acidity of $SiCl_4$. This favors associative nucleophilic substitution on silicon, which occurs very rapidly. In contrast, CCl_4 behaves as a Lewis base and nucleophilic substitution at the carbon is possible only according to a dissociative mechanism, which is much slower. Due to the small size of the carbon atom, a concerted mechanism (S_N2) is unlikely.

A similar example can be found with complexes. $[Co(NH_3)_6]^{3+}$ is thermodynamically stable with respect to decomposition into Co^{3+} and 6 NH_3, but unstable upon acid hydrolysis, forming $[Co(H_2O)_6]^{3+}$. However, the decomposition of the complex takes several days, i.e. kinetically the complex is relatively stable with respect to acid hydrolysis. The reason for this is that there is no favorable pathway for this reaction. Substitution reactions on complexes can proceed according to an S_N1 or S_N2 mechanism. In the former, the rate-determining step is the release of a ligand, thus creating a vacant coordination site for a new ligand (dissociative mechanism). The second is that another ligand is first added while increasing the coordination number (rate-determining), and then a ligand leaves to return to the original coordination number (associative mechanism). Ligand exchange at the cobalt requires a change in the coordination number at the metal center. In the first, the coordination number at the cobalt is temporarily decreased to five; in the second, it is temporarily increased to seven. As a d^6 ion in the low spin state, the cobalt(III) ion has a particularly high crystal field stabilization energy, which must be overcome to some extent. The activation energy is very high and the reaction is very slow.

$$[Co(NH_3)_6]^{3+} + 6\,H_3O^+ \rightarrow [Co(H_2O)_6]^{3+} + 6\,NH_4^+$$

The influence of the crystal field stabilization energy on the kinetic stability of the complexes is also shown by the following series. Cyanido complexes are thermodynamically very stable, but they can have a different kinetic stability. Whether the complexes are kinetically labile (rapid ligand exchange with other cyanide ligands) or kinetically stable (very slow ligand exchange) can be detected by the incorporation of labeled cyanide (e.g., ^{14}C, radioactive, or ^{13}C or ^{15}N, for NMR experiments). Thus, the half-lives for CN^- exchange were determined for $[Ni(CN)_4]^{2-}$ (30 s), $[Mn(CN)_6]^{3-}$ (1 h), and $[Cr(CN)_6]^{3-}$ (24 days). The very short half-life of the nickel complex can be explained by the possibility of increasing the coordination number from 4 to 5. Thus, an addition-elimination mechanism takes place here. In the case of the octahedral chromium(III) complex, a d^3 ion, the high kinetic stability can again be explained by the high crystal field stabilization energy. Here, the energetically lower t_{2g} orbitals are half occupied and the deviation from the ideal octahedral geometry leads to energy loss. The corresponding manganese complex, being a d^4 ion, already has a distorted geometry (see Jahn-Teller effect) and the ligand exchange proceeds faster. The kinetic stability may (but need not) provide clues to the reaction mechanism and activation energy.

6.3 The Chelate Effect

Complexes with multidentate ligands (so-called chelate ligands with the resulting chelate complexes) are more stable than complexes with comparable monodentate ligands. This phenomenon, known as the chelate effect, will be examined in more detail below.

As an example, we consider the reaction of Cd^{2+} with the monodentate methylamine or the bidentate ethylenediamine as ligands. The equilibrium constant for reaction (b) with ethylenediamine is significantly larger than for (a) with the monodentate ligands.

$$[Cd(H_2O)_6]^{2+} + 4\,NH_2Me \rightleftharpoons [Cd(NH_2Me)_4(H_2O)_2]^{2+} + 4\,H_2O \quad (a)$$
$$[Cd(H_2O)_6]^{2+} + 2\,en \rightleftharpoons [Cd(en)_2(H_2O)_2]^{2+} + 4\,H_2O \quad (b)$$
$$K(b) \ll K \quad (a)$$

The stability of chelate complexes can be explained by both kinetic (inertness) and thermodynamic aspects. For kinetic stability, we can imagine that for chelate ligands, the coordination of the first donor atom at the metal center causes the second (and all others) to come into spatial proximity with the metal center and thus to coordinate faster. If we stick with the scheme shown in Fig. 6.3, it means that the activation energy for the coordination of the second donor atom is lowered. This reasoning also slows down ligand exchange because both (or all) bonds between the metal center and the ligand must be broken before the ligand can dissociate. The stability of the complex increases.

For thermodynamically stable complexes, the free energy ΔG^0 of the reaction is negative. If one uses the *Gibbs-Helmholtz equation*

$$\Delta G^0 = \Delta H^0 - T\,\Delta S^0$$

then it can be seen that two factors play a role in achieving this condition. The first is $\Delta H^0 < 0$, which corresponds to an exothermic complexation reaction, while for the second, $\Delta S^0 > 0$, there is an increase in entropy during complexation. The ligands methylamine and ethylenediamine given in the example reaction are very similar in terms of their position in the spectrochemical series, so ΔH^0 is initially comparable in both cases. Comparing the equations (a) and (b), we find that in (a) the particle number is the same on both sides of the equilibrium, whereas in (b) the particle number is increased from 3 to 5 upon complexation with a chelating ligand. This increases the entropy (disorder) of the system, which is associated with an (additional) energy gain ($\Delta S^0 > 0$). Due to the increase in entropy, chelate complexes are thermodynamically more stable than analogous complexes with monodentate ligands.

In Fig. 6.4, the equilibrium constants for the formation of copper complexes with four monodentate (NH_3), two bidentate (en = 1,2-diaminoethane), one open-chain tetradentate (trien = triethylenetetraamine), and one macrocyclic tetradentate (taa = tetraaza[12] annulene) ligands are given. We see that as the number of donor atoms per ligand increases, the complex formation constant increases. The complex formed is more and more stable and our previous reasoning is confirmed. Comparing the increase in equilibrium constants, it is noticeable that from ammonia (NH_3) to ethylenediamine (en), the increase in stability is significantly greater than for the second step (en→ triene). This is due to the fact that enthalpy cannot be ignored in this ranking. The alkylation of the nitrogen ligands slightly increases its basicity, which is reflected in the ligand field splitting and further increases the stability of the chelate complex. Another aspect is the down-regulation of the steric ligand-ligand interaction. In the complex with four ammonia ligands, repulsive interactions take place between the ligands. By introducing a covalent bridge, these interactions, which lower the complex stability, are suppressed. This effect also plays a role in the cadmium complexes discussed above. Both are pure enthalpy effects, which also contribute to the stability of chelate complexes.

If we compare the two complexes with tetradentate ligands, $[Cu(trien)]^{2+}$ and $[Cu(taa)]^{2+}$, we find that for the macrocyclic ligand taa the complex stability is further increased. The increased stability of a complex with macrocyclic ligands over a comparable

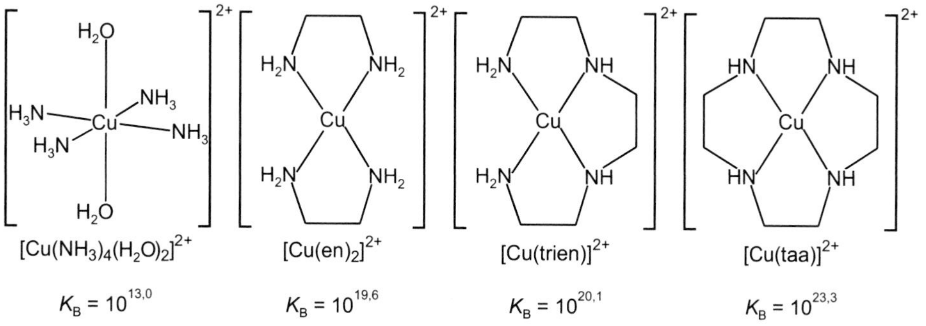

Fig. 6.4 Equilibrium constants for the reaction of $[Cu(H_2O)_6]^{2+}$ with 4 NH_3, 2 en, 1 triene and taa = tetraaza[12]annulene, respectively. For reasons of clarity, only for $[Cu(NH_3)_4(H_2O)_2]^{2+}$ the possibly still coordinated water ligands have been drawn in

complex with open-chain ligands is called the *macrocyclic effect*. In this case, the number of particles mentioned at the beginning does not play a role and an important contribution to the increased stability of the system is provided mainly by the preorganization of the ligand and the repulsive interactions already discussed, which are absent in macrocyclic ligands. In a macrocyclic ligand, all donor atoms are already optimally arranged for coordination of the metal center. An open-chain ligand tends to have a linear structure in solution, where the donor atoms can point in different directions. When coordinating to the metal center, the ligand must reorient itself, which requires energy. In solution, the free open-chain ligand has more degrees of freedom, e.g. with respect to rotation about individual bonds, than the macrocyclic ligand, which is already pre-oriented for complexation. Upon complexation, the number of degrees of freedom for the open-chain ligand decreases, and the entropy decreases. Upon complexation of the macrocyclic ligand, this entropy decrease does not occur and the complex is more stable. We see that the entropy of a system can depend not only on the number of particles, but also on the "disorder" of a ligand.

Different ring sizes also affect the stability of a chelate complex, with the preferred ring size depending on the size of the metal centre. An example of this is given in Fig. 6.5, where the stability of a copper complex with two ethylenediamine and two propylenediamine ligands, respectively, is compared. In the case of ethylenediamine, a five-membered chelate ring is formed, whereas propylenediamine leads to the formation of six-membered chelate rings. In the case of copper complexes, five-membered rings are preferred, which then coordinate largely stress-free at the metal center.

The chelate effect is exploited in the complexometric titration with H_2edta^{2-}. H_2edta^{2-} is a six-dentate ligand that forms very stable complexes with a variety of metal ions. Because of the four acid groups, the stability of the complexes is pH-dependent, at the metal center coordinates the four-fold deprotonated anion $edta^{4-}$. Its pH-dependent concentration in solution is given by the pK_S values: $pK_S = 1.99$; 2.67; 6.16 and 10.26. Figure 6.6 gives the corresponding chemical equation. This pH dependence affects the

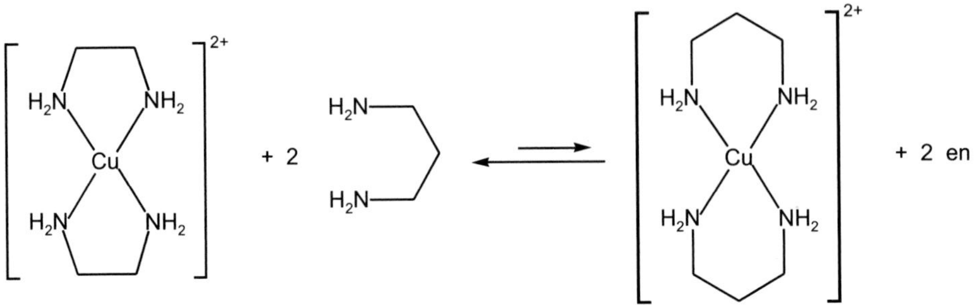

Fig. 6.5 Comparison of the stability of 5-membered rings and 6-membered rings in copper complexes. Preferred are five-membered rings which are largely stress-free, planar or slightly corrugated in construction

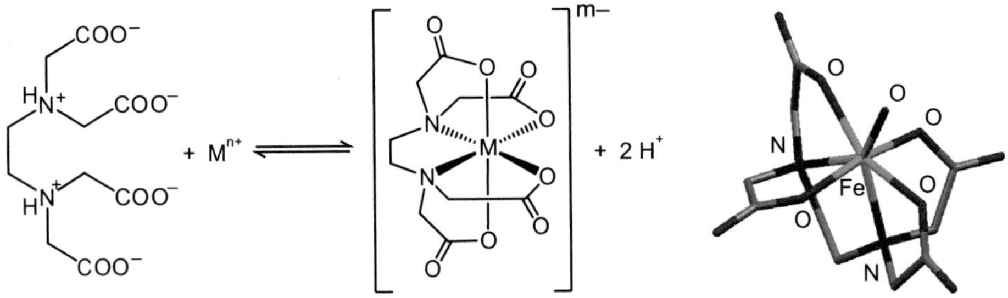

Fig. 6.6 Complexation of a metal ion with H_2edta^{2-}. Protons are released during the reaction. From the crystal structure of an iron(III) edta complex shown on the right, it can be seen that the iron ion has coordination number 7 and there is also an additional water acting as a ligand [21]

complex formation. Thus, hard cations such as Fe^{3+} and Al^{3+} form stable complexes even in neutral solutions, whereas softer cations require higher pH values to form stable complexes. To explain this difference, we can again use the HSAB concept. With four oxygen and two nitrogen atoms as donor atoms, $edta^{4-}$ is a hard Lewis base and forms more stable complexes with hard Lewis acids.

$edta^{4-}$ is not suitable for complexing alkali cations with a single positive charge. For this purpose, crown ethers were developed, which are discussed in more detail in the chapter Supramolecular Chemistry.

6.3.1 Chelation Therapy

The use of multidentate ligands to remove metal ions from the organism is the basis of chelation therapy in cases of heavy metal poisoning or metal ion accumulation as a result of metabolic disorders ("Only the dose makes the poison."). For example, in cases of lead poisoning, attempts are made to replace the consumption of large quantities of butter as an

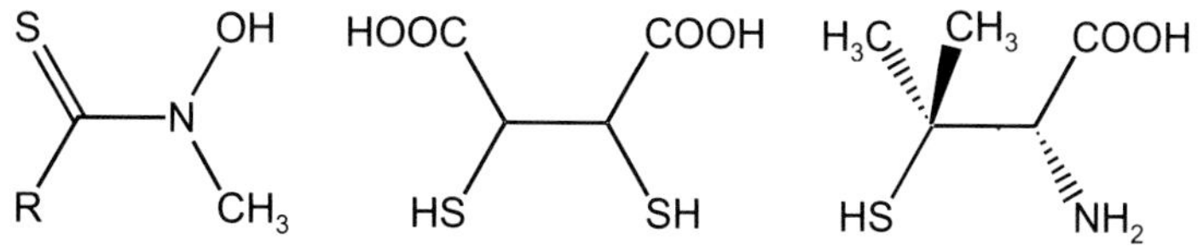

Fig. 6.7 Ligands for chelation therapy in heavy metal poisoning. Left: Thiohydroxamate-containing S,O-ligand for lead poisoning, Middle: S,S-ligand dimercaptosuccinic acid, Right: N,S-ligand D-penicillamine for the therapy of Wilson's disease

old household remedy with the application of ligands containing thiohydroxamate, while mercury poisoning is treated with dimercaptosuccinic acid. The chelating ligands used are always matched to the metal ion to be removed. For example, Cd^{2+} and Cu^{2+} prefer N,S ligands, while ligands with exclusive S coordination are more suitable for arsenic and mercury poisonings. These differences can again be explained by the HSAB theory. As the ionic radius increases from Cu^{2+} to Cd^{2+} to Pb^{2+}, the ions are more and more soft. Accordingly, the use of soft donor atoms in the case of lead leads to stable complexes, while the slightly harder metal centers also prefer harder donor atoms (N instead of S). For Cd^{2+} and Pb^{2+} S-containing polychelate ligands have also proved successful. Since the complexes formed must be stable in the physiological pH range and excretable in urine, additional hydrophilic groups (-OH, -COOH) are used. All these ligands have only a low selectivity and lead to side effects, which is why they are only used as an emergency measure. Selected examples are given in Fig. 6.7.

The disorder of the biosynthesis of the copper-binding serum protein coeruloplasmin, known as *Wilson's disease,* leads to toxic copper accumulation in the tissues of the affected organs. Here, chelation therapy with D-penicillamine has proven effective.

In certain blood diseases (e.g. sickle cell anaemia), regular blood transfusions lead to an accumulation of iron in the body, which puts a strain on various organs and can even lead to death. Again, attempts to remove excess iron from the body are based on chelation therapy. To date, the siderophore desferrioxamine B, a linear peptide, has been administered for this purpose. Siderophores are low molecular weight iron transport molecules. Iron is an essential element for almost all organisms. While we can cover our iron requirements through food (oral uptake), this is not possible for plants and microorganisms in this way. Other (easily soluble) ions can simply be absorbed with the water. The problem with iron is that it is extremely difficult to dissolve in the most common oxidation state +3 (solubility product $Fe(OH)_3$: 5.0×10^{-38} mol^4/l^4). Especially in calcareous (basic) soils, plants and microorganisms face the problem of mobilizing iron for uptake from the soil. They achieve this by using siderophores, which are extremely strong complexing agents for iron(III). These form very stable water-soluble complexes with the iron under the alkaline conditions in the soil, which can now be taken up by the plant (then we speak of phyto-siderophores) or the microorganisms. In a next step, the iron has to be released again to be available for incorporation, e.g. into electron transfer proteins or catalytic centres. The plant or microorganism achieves this by having a lower pH in the tissue than in the surrounding soil. This

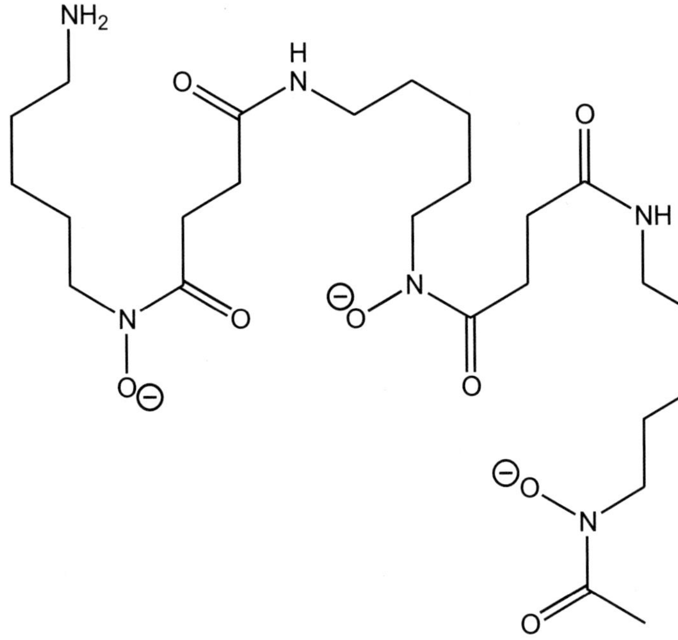

Fig. 6.8 Structure of desferrioxamine B

leads to a (partial) protonation of the siderophore, which is accompanied by a decrease in complex stability. The iron is now available for further complexation reactions. Figure 6.8 shows the structure of the siderophore desferrioxamine B.

6.3.2   Radiotherapy and MRI

Another clinical field of application of chelate complexes with increasing importance are radiopharmaceuticals which are mainly used for diagnostic purposes. The radioactive nuclides used for this purpose are mostly gamma emitters of relatively low energy, which can be easily detected with scintillation counters. They are used for visualization (imaging) of diseased organs in which the radioactive compounds preferentially accumulate. Technetium compounds are used particularly frequently. The enrichment of the complexes in the target regions is achieved by receptor-specific molecules. Due to its radioactive properties, only a short residence time is desired for the technetium in the body. This is achieved by complexation with chelating ligands. The resulting complexes must be very stable and readily soluble in water in order to be readily excreted. Examples of approved technetium-based radiotherapeutics are given in Fig. 6.9.

Magnetic resonance imaging (MRI) is based on the principles of nuclear magnetic resonance (like NMR spectroscopy) and is an imaging technique that can be used to

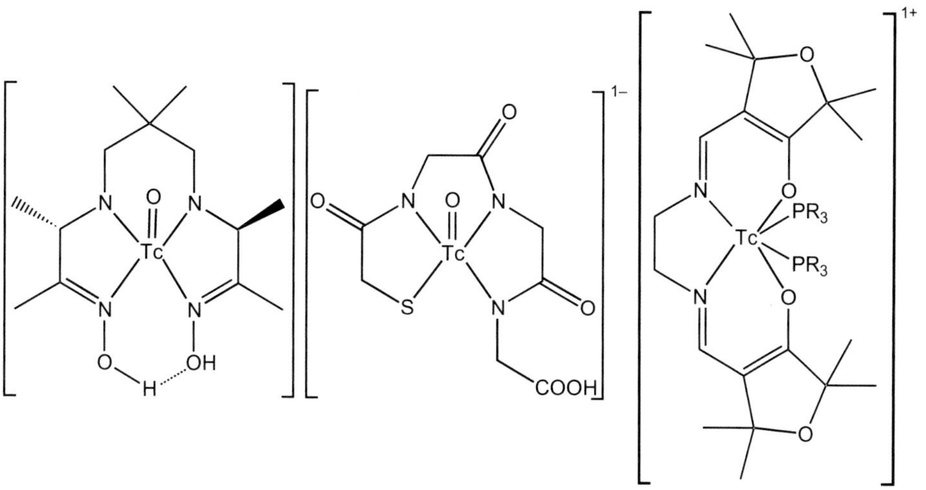

Fig. 6.9 Examples of currently approved radiotherapeutics based on technetium

visualize tissue and organs and produce sectional images. A strong external magnetic field is used to split the nuclear spins of protons (e.g. from water in the body or from tissue) into two energy levels between which the transition is excited. The contrast in the images is due to the different relaxation times of the protons in the different types of tissue. In addition, the different content of protons in the different tissues also plays a role. To further increase the contrast, contrast agents are added. Among others, gadolinium(III) chelate complexes are used here, which lead to a shortening of the relaxation time in the vicinity of the paramagnetic centers due to the paramagnetic properties of gadolinium. Due to the toxic properties of the gadolinium(III) ion, it must be incorporated into very stable complexes that are stable in the body and are completely excreted. An example of an approved gadolinium complex is given in Fig. 6.10. Current research is concerned with the search for so-called "intelligent" contrast agents. These could be, for example, spin crossover systems that respond to differences between normal tissue and tumor tissue (e.g., as function of pH) with a spin transition, leading to differences in contrast that ensure better detection of tumor tissue [22].

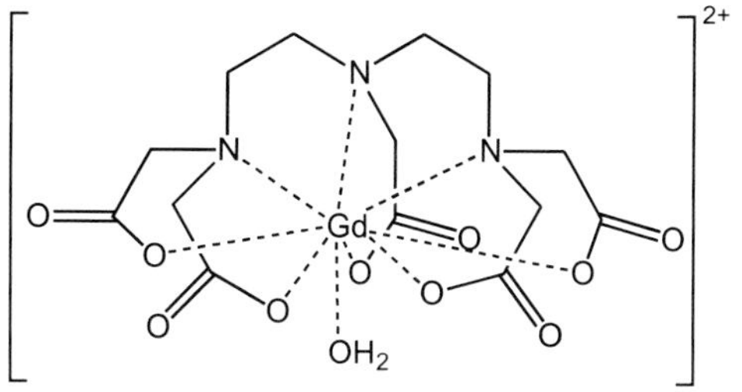

Fig. 6.10 Structure of the first gadolinium(III) complex [Gd(DTPA)(H$_2$O)]$^{2-}$ approved as a contrast agent for MRI. (Magnevist 1988)

6.4   The *Trans* Effect

So far, we have looked at the stability of complexes from different aspects. We have always considered the complex as a whole and, in the case of ligand exchange reactions, discussed the exchange of all ligands. In square-planar heteroleptic complexes, there is an effect that selectively weakens certain bonds, leading to preferential (faster) substitution of one or more ligands. Heteroleptic complexes are complexes that have at least two different ligands. The opposite of this is homoleptic complexes, where all the ligands are the same. The kinetic effect to be discussed now is called the *trans effect* and is shown schematically in Fig. 6.11.

In the complex [MABX$_2$], if one of the ligands X is substituted by Y to form the complex [MABXY], then the ligand Y can enter in *trans position* to A or B. It has been shown that the ligands have varying degrees of ability to direct newly entering substituents to the *trans position.*

For practical application of this effect in substitution reactions, the strength of the *trans effect* of each ligand must be known. In the following, the sequence of some important ligands is given:

$$CN^- > CO > C_2H_4 \approx NO > PR_3 > SR_2 > NO_2^- > SCN^- > I^- > Cl^- > NH_3$$
$$> py > RNH_2 > OH^- > H_2O$$

6.4.1   Interpretation of the *Trans* Effect

There are various models for the interpretation of the *trans effect*. Most of them assume that in square-planar complexes the two *trans* ligands compete for a common p- or d-orbital of

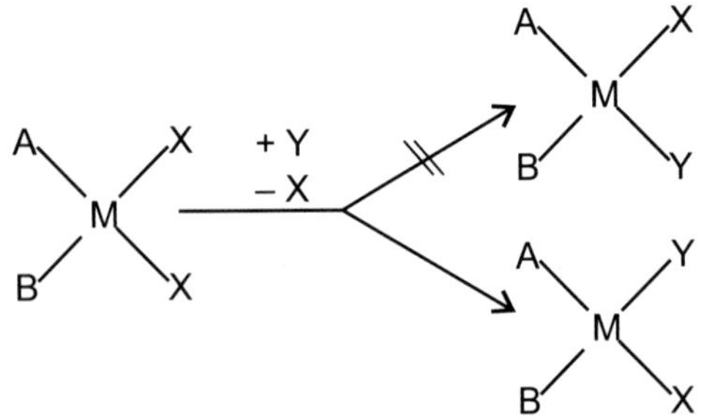

Fig. 6.11 Schematic representation of the *trans effect*. If the *trans effect* of ligand B is greater than that of ligand A, the substitution reaction takes place preferentially in the position *trans* to B

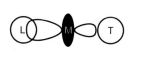

Fig. 6.12 Interpretation of the *trans effect*. In square planar complexes, the two *trans* ligands compete for a common orbital at the metal center. If the bond to one ligand is particularly strong, the bond to the *trans* ligand is weakened

the metal center. For ligands with a high inclination to form π bonds (CO, NO, C_2H_4), the formation of a strong π bond activates the bond of the central ion to the *trans*-ligand for substitution (theory of *Chatt* and *Orgel*). This means that the bond to the *trans*-ligand is weakened and the ligand is more easily cleaved off. The reaction rate for the substitution of the *trans*-ligand is increased and his substitution occurs preferentially because of this. This discussion implies that ligand substitution proceeds via a dissociative mechanism in which one ligand is cleaved off first, and then the new ligand is attached. This model fails for ligands that are not capable of forming π bonds, but where a *trans effect* is still observed (e.g. NH_3). Here, it is discussed that the *trans* ligands enter into competition for a p orbital. That is, when one ligand forms a strong bond, the one *trans* to the ligand is weakened and the latter can be cleaved more easily – or, more precisely, more quickly. In Fig. 6.12, the competing interaction of the two ligands for the common orbital at the metal center is again shown schematically. Regardless of whether the bonds are π or σ, the basic principle is the same in both models. The strong bond to a ligand corresponds to a large overlap between the d or p orbital of the metal and the ligand orbital. As a result, the electron density of the orbital on the metal is shifted towards the ligand with strong *trans effect* (larger orbital lobes = larger residence probability for the electron). As a result, the orbital lobes on the *trans side* are smaller and can only form a weak bond (small overlap of the orbitals), which can be easily dissolved.

6.4.2   Cisplatin and the *Trans* Effect

An important application of coordination chemistry for therapeutics is the use of platinum complexes in cancer therapy. The cytostatic activity of *cis*-diamminedichloridoplatin (II) ("cisplatin") was discovered in the early 1960s. The compound has been clinically approved for cancer therapy since 1978. In the meantime, a number of other Pt (II) complexes have been synthesized and tested for their cytostatic activity. Primarily – but not exclusively! – the *cis*-isomers show cytostatic activity.

The effectiveness of the platinum complexes is due to their binding of the{Pt $(NH_3)_2\}^{2+}$ fragment to the nitrogen atoms of the DNA nucleotide bases, with the N7 position of guanine appearing to be particularly favoured, as shown in Fig. 6.13. The

Fig. 6.13 Coordination of
cisplatin to a DNA strand. The
$\{Pt(NH_3)_2\}^{2+}$ fragment has been
circled

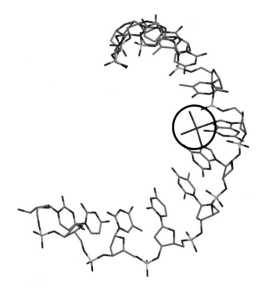

resulting structural change in DNA leads, among other things, to inhibition of DNA synthesis at a single-stranded template. In tumor cells, the repair mechanisms for DNA are partially disabled, so the modification of DNA by Pt binding is not reversed. In addition, tumor cells have a higher growth rate. For this reason, the cytostatic effect is higher in tumor cells than in normal cells and the compounds are successful in cancer therapy.

The chemist is now faced with the question of how he or she can specifically produce the *cis*-isomer (or possibly also the *trans*-compound) of platinum(II) complexes. This is where the *trans effect* comes in handy. Comparing the directing effect of Cl$^-$ and NH$_3$ on [PtCl(NH$_3$)$_3$]$^+$, it can be seen that the Cl$^-$ ion has a larger *trans* effect than the NH$_3$ molecule. The reaction of this complex with Cl$^-$ gives *trans*-[PtCl$_2$(NH$_3$)$_2$]. On the other hand, if we start from [PtCl$_3$(NH$_3$)]$^-$ and exchange a chloride for ammonia, we obtain *cis*-[PtCl$_2$(NH$_3$)$_2$]. We see: the *trans effect* is most important for the synthesis of *cis*- or *trans*-complexes. In Fig. 6.14 the two reactions are shown again.

Since ammonia and chloride are close to each other in the ranking for the strength of the *trans effect*, the large-scale preparation is carried out via the iodide. Starting from the tetrachloridoplatinate(II), the tetraiodidoplatinate(II) is prepared by reaction with an excess of potassium iodide. Subsequent reaction with ammonia yields the *cis*-diammine diiodidoplatinate(II). The next step is to add silver nitrate to the aqueous solution, precipitating the poorly soluble silver iodide and leaving the *cis*-diamminequaplatin(II) ion in solution. Addition of potassium chloride precipitates the product *cis*-diamminedichloridoplatin(II).

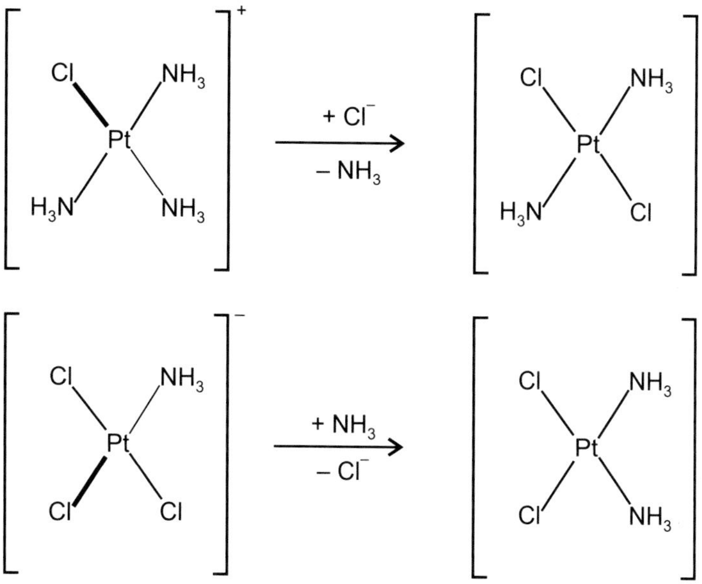

Fig. 6.14 Exploitation of the *trans effect* for the selective synthesis of *cis-* or *trans*-platinum complexes. The different strength of the bonds is illustrated by different line thicknesses

6.5   Questions

1. Why does edta^{4-} form stable complexes with hard cations even at neutral pH values, while higher pH values are required for soft cations?
2. In the complexometric determination of ferric ions, sulfosalicylic acid is added as an indicator and titrated with edta^{4-}. The color change is from red to colorless. The colour change is from red to colourless. What happened?
3. Why does this detection not work quantitatively for iron(II) ions? Estimate the thermodynamic and kinetic stability of the complexes involved.
4. What is the difference between a thermodynamically stable complex and a kinetically stable complex? Can both attributes be combined in one complex?
5. Explain the chelating effect observed for the complexation of multidentate ligands! How and under what conditions can the effect be used analytically?
6. Write down the reaction equations for the uptake of iron(III) from the soil with the help of a siderophore. Which equilibria compete with each other?
7. Cisplatin (*cis*-diamminedichloridoplatin(II)) is an important drug in cancer therapy. How can this complex be specifically produced and what effect does it exploit?

Redox Reactions of Coordination Compounds

Chemical reactions can be divided into two categories, substitution reactions and redox reactions. In the previous chapter, we considered the stability of complexes and dealt with ligand exchange, i.e. substitution reactions. In the following, we will deal with the second major class of reactions of complexes, the redox reactions.

Understanding electron transfer between complexes is essential for understanding life as we know it and the chemical processes involved. Whether photosynthesis or respiration – all these reactions are redox reactions and extremely fast electron transfer is essential for these processes to take place. Many of the proteins involved in these reactions are metalloproteins in which one or more metal ions are bound to the protein with the help of side chains that act as ligands. Thus, a complex is present. In this context, it is not surprising that two Nobel Prizes have already been awarded for investigations into the theory of electron transfer in chemical systems. In 1983, the Nobel Prize in Chemistry was awarded to Henry Taube for his work on the mechanism of electron transfer reactions, particularly in complexes. [23] Nine years later, in 1992, Rudolph A. Marcus was awarded for his contribution to the theory of electron transfer reactions in chemical systems. [24] The Marcus theory he established allows the description of a variety of different phenomena such as photosynthesis, corrosion, chemiluminescence, or the conductivity of electrically conducting polymers.

To get started, let's look at electron transfer using blue copper proteins as an example.

B. Weber, *Coordination Chemistry*, https://doi.org/10.1007/978-3-662-66441-4_7

7.1   Blue Copper Proteins

Copper frequently occurs in electron transfer chains in biological systems in the form of blue copper proteins. In these metalloproteins, the copper atom is located in a similar coordination sphere in each case, in which it can be present in the oxidation states +1 and + 2. It is surrounded by four monodentate ligands: two histidines (N), one cysteine (S), and one methionine (S) and thus has a N_2S_2 -coordination environment. Cu^{1+} is a d^{10} ion. For coordination number 4, the preferred coordination environment is a tetrahedron with all four ligands as far apart as possible. Due to the closed d shell, different ligand fields do not play a role here; no other coordination geometry is preferred as there is no crystal field stabilization energy. In contrast, for the d^9 ion Cu^{2+} the preferred coordination environments are square planar, square pyramidal, or a Jahn-Teller distorted octahedron. That is, a redox reaction is expected to change the coordination environment. Such structural changes would significantly hinder and slow down electron transfer. To circumvent this, in the blue copper proteins the copper is in a coordination environment intermediate between that of a tetrahedron and a square planar coordination environment. In Fig. 7.1, a section of the structure of the active site of the blue copper protein plastocyanin is given. This distorted coordination environment is enforced by the protein regardless of the oxidation state of the copper. In this way, a very fast electron transfer can be realized as the activation energy is strongly lowered. This strained state, which lowers the activation barrier of a process mediated by a protein (enzyme), is called the *entatic state*.

Before considering in more detail how the distorted coordination environment imposed by the protein lowers the activation energy for electron transfer, we clarify the question of why copper(II) ions do not occur with an ideal octahedral coordination environment.

7.1.1   The Jahn-Teller Effect

Why are copper(II) complexes with six ligands not ideally octahedral? Deviations from highly symmetric molecular geometries are not only observed in heteroleptic complexes (= complexes with different ligands), but sometimes also when all ligands of a complex are identical. Such a complex is called a homoleptic complex. In principle, a distorted coordination geometry is always observed when the electronic ground state in the ligand field is

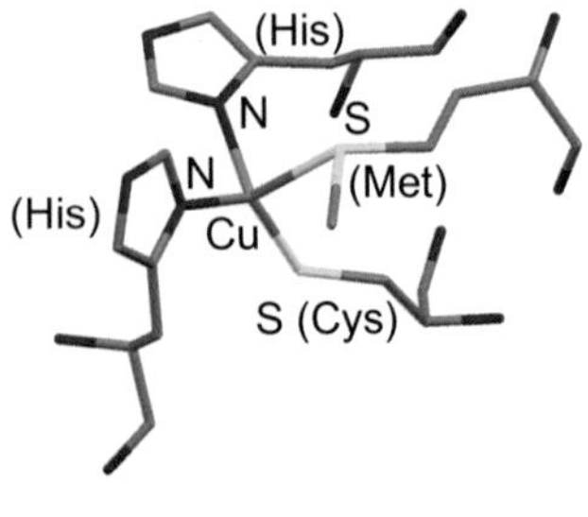

Fig. 7.1 Structure of the active site of the blue copper protein plastocyanin. The four ligands are histidine 37, cysteine 84, histidine 87, and methionine 92

Fig. 7.2 Microstates for the ground state of a copper(II) ion in the octahedral ligand field

degenerate (i.e., there are several microstates, see term symbols). Then the following applies:

> **Jahn-Teller Theorem**
>
> In any nonlinear molecule whose electronic ground state is degenerate, there is a natural oscillation that cancels the degeneracy of the ground state. The ground state is stabilized in a geometry distorted by this oscillation.

To better illustrate this issue, we consider the two possible microstates for the ground state of a copper(II) ion (d^9) in the octahedral ligand field. The corresponding term is ^2E_g and the two associated microstates are given in Fig. 7.2.

A natural oscillation that cancels the degeneracy of this ground state corresponds to the stretching and compression of the octahedral ligand field along the z-axis. This removes the degeneracy of the two e_g orbitals d_{z^2} and $d_{x^2-y^2}$. For a stretched octahedron, the d_{z^2} orbital is slightly lowered in energy and the $d_{x^2-y^2}$-orbital is raised by the same amount of energy; for the compressed octahedron, it is the other way around. This distortion removes the degeneracy of the ground state with energy gain for the system. In a d^9 system, of course, the t_{2g} orbitals are also energetically split during this oscillation. However, this splitting does not contribute to the energy gain, since all orbitals are fully occupied. However, it can be detected by observing a shoulder at the band of the d-d transition in the optical spectrum of such a complex. Figure 7.3 illustrates the described concept once again.

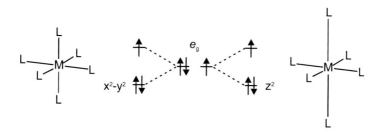

Fig. 7.3 Splitting of the e_g orbitals of a d^9 ion as a function of the distortion (left compressed, right stretched) of the octahedron

Based on the so far made observation, nothing can be said about the nature and magnitude of the Jahn-Teller effect. In general, one observes that the distortion due to an incompletely filled t_{2g} shell is smaller than those originating from e_g^1 and e_g^3 configurations. Most of the distorted octahedra studied are stretched square bipyramids.

The Jahn-Teller effect not only occurs in octahedral complexes, but can be transferred to other systems. An example of a pronounced Jahn-Teller effect at coordination number four are square planar high-spin iron(II) complexes with diolato ligands. As we have already learned, a square planar coordination environment is expected mainly for d^8 transition metal complexes with large ligand field splitting, while a tetrahedral coordination geometry is typical for small ligand field splitting and other d electron configurations. However, for tetrahedral d^6 high-spin complexes, in some cases planarization of the ligand field leads to a more stable coordination environment, as shown on the left [95]. Also, the distorted structure of cyclobutadiene can be explained by a Jahn-Teller-like distortion leading to the localization of the double bonds. The corresponding Walsh diagram is shown on the right.

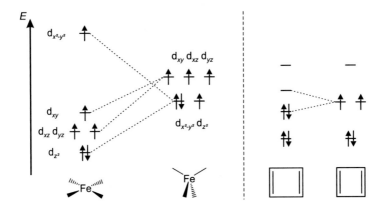

7.2   Redox Reactions in Coordination Compounds

Redox reactions are probably the mechanistically best-studied chemical processes of complexes. A redox reaction between complexes need not be associated with a material change in the overall balance. However, in many cases the redox transformation cannot be described by the electron transfer step alone, but consists of a complex sequence of ion pair formation, ligand substitution or ligand transfer between the molecules involved. In Fig. 7.4, three examples of an electron transfer reaction in coordination chemistry are given.

a $[^*Co(NH_3)_6]^{2+}$ + $[Co(NH_3)_6]^{3+}$ $\rightleftharpoons$ $[^*Co(NH_3)_6]^{3+}$ + $[Co(NH_3)_6]^{2+}$

b $[CoCl(NH_3)_5]^{2+}$ + $[Cr(H_2O)_6]^{2+}$ + $5H^+$ + $5H_2O$ $\longrightarrow$ $[Co(H_2O)_6]^{2+}$ + $[CrCl(H_2O)_5]^{2+}$ + $5NH_4^+$

c $[Fe(H_2O)_6]^{2+}$ + $[Ru(bipy)_3]^{3+}$ $\longrightarrow$ $[Fe(H_2O)_6]^{3+}$ + $[Ru(bipy)_3]^{2+}$

Fig. 7.4 Examples of redox reactions in coordination chemistry (**a**) Self-exchange redox process, (**b**) Inner sphere mechanism, (**c**) Outer sphere mechanism

Equation (a) is an example of electron transfer between the same corresponding redox pairs. Such self-exchange processes were studied in the 1950s using isotopically labeled metal complexes. The results of these investigations led to the Marcus theory of electron transfer in chemical reactions established by *Rudolph A.* Marcus (Nobel Prize in Chemistry 1992), which will be described in detail below.

In the case of different complex types (eqs. b and c), the product distribution sometimes allows conclusions to be drawn about the reaction mechanism. This depends strongly on the redox partners and their reactivity towards ligand substitutions (kinetically inert or labile, see stability of complexes). Equation b) shows the reduction of a substitution-inert amminchloridocobalt(III) complexes by the substitution-labile hexaaquachromium (II) complex $[Cr(H_2O)_6]^{2+}$. Since the chlorido ligand was found in the inert Cr(III) product, it is assumed that at the time of electron transfer this ligand must be coordinated to both metal centers and it acts as a bridging ligand ("conducting link") during electron transfer. Since the reactants in this mechanism simultaneously share a ligand in the inner coordination sphere, it is called the inner sphere mechanism. Electron transfer between bridged metal centers was studied by *Henry Taube*, who was awarded the Nobel Prize in Chemistry for it in 1983. In reaction (c), the coordination spheres of the complexes involved remain intact. The mechanism is comparable to the self-exchange process shown in equation (a) and can also be explained by Marcus theory. The electron transfer takes place without the mediation of a common bridging ligand, which is why this mechanism is called the outer-sphere mechanism.

7.2.1   The Outer Sphere Mechanism

Redox reactions in complexes take place according to the outer-sphere mechanism if at least one of the complexes involved is kinetically inert and/or the ligands are not suitable for bridging metal centers. Under these conditions, the formation of an inner sphere complex in which electron transfer occurs via the bridging ligand is not possible. Self-exchange processes are the simplest variant for redox reactions by this mechanism. For a number of complexes, isotopically labelled compounds were used to determine the rate constants, which are given in Table 7.1.

Table 7.1 Rate constants (k) for self-exchange of selected redox pairs and change in metal-ligand distance Δr during electron transfer

Corresponding redox pair	k (M^{-1} s^{-1})	Δr (pm)
$[Co(NH_3)_6]^{2+/3+}$	$8 \cdot 10^{-6}$	17.8
$[V(H_2O)_6]^{2+/3+}$	$1.0 \cdot 10^{-2}$	
$[Fe(H_2O)_6]^{2+/3+}$	4.2	13
$[Co(bipy)_3]^{2+/3+}$	18	
$[Ru(NH_3)_6]^{2+/3+}$	$3 \cdot 10^3$	4
$[Ir(Cl)_6]^{3-/2-}$	$2.3 \cdot 10^5$	
$[Ru(bipy)_3]^{2+/3+}$	$4.2 \cdot 10^8$	≈ 0

For self-exchange reactions, there is no thermodynamic driving force that affects the reaction rate. From this point of view, the wide variation of the given rate constants is impressive. The data in Table 7.1 already indicate that the rate of electron transfer depends on the extent of change in the metal-ligand distance. The pioneering work in developing a powerful theory for estimating the rate of reaction can be attributed to *Rudolph A. Marcus*. He was able to show that three main factors influence the reaction rate of the overall process.

1. The approach of the reactants and the **formation of the outer sphere complex**, in which the electronic interaction between the associated reactants provides the prerequisite for the "delocalization" of an electron from one center to the other.
2. The barrier of the electron transfer step resulting from the differences in the equilibrium structures of the reduced and oxidized species. (**Vibrational barrier**)
3. The barrier to charge redistribution caused by the surrounding solvent of the outer sphere complex. (**Solvent barrier**)

Formation of the Outer Sphere Complex Electron transfer between two metal centres can only take place in the contact state (outer sphere complex) with a fixed intermolecular distance. The probability for the formation of this aggregate and its stability (lifetime) strongly depends on the reaction partners, especially their charge. Whereas in the case of equally charged complexes, as in the case of self-exchange processes, a short-lived "encounter complex" is formed which immediately decays again due to the repulsive interactions, in the case of oppositely charged complexes a supramolecular aggregate with directly determinable stability can be formed. An example for this would be an electron transfer between $[Co(NH_3)_5(py)]^{3+}$ and $[Fe(CN)_6]^{4-}$. The electrostatic attraction between the two reactants increases the stability of the outer-sphere complex and thus the reaction rate.

The Vibrational Barrier The second factor is the so-called "vibrational barrier". In Table 7.1 we already see that the rate of electron transfer depends strongly on the change of the metal-ligand distance. A large change of the metal-ligand distance during the change of the oxidation state of the metal center leads to a much slower electron transfer. The

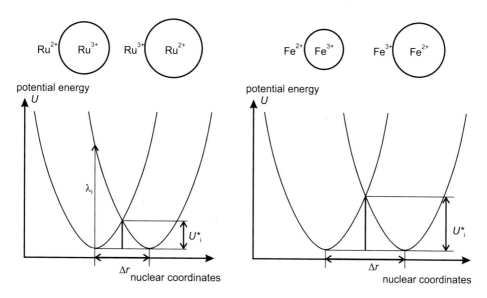

Fig. 7.5 Schematic representation of the influence of the change in the M-L distance (Δr) on the energy U_i^* for electron transfer in the self-exchange process. Shown is the potential energy U as a function of the metal-ligand distance for one of the two reactants. A second set of such potential curves exists for the second reactant. Electron transfer can only take place at the point of intersection of the two potential curves, since significantly more energy is required for a vertical transition without alignment of the structures (λ_i)

charge of the complexes studied in the self-exchange processes is always the same and does not affect the different rates.

As a first example, the self-exchange reaction of the redox couple $[Fe(H_2O)_6]^{2+/3+}$ is used. The most important structural difference between the divalent and trivalent forms is the Fe-OH$_2$ bond length, which shortens from 2.10 to 1.97 Å upon oxidation. In both cases, the iron is in the high-spin state, but the ionic radius of the high-spin iron(III) ion with a half-occupied d shell is smaller than that of the high-spin iron(II) ion, which still has an extra electron in the 3d shell. In a redox reaction, the bond lengths of the two reactants must change. Because of the much lower mass of electrons compared to atoms, they move much faster. To get a feeling for the orders of magnitude, let's compare the following numbers. The mean collision time of molecules in a solution is 10^{-10} s, a molecular vibration is about 10^{-13} s long and an electron excitation takes 10^{-15} s. This means that during the average collision time, a molecule vibrates 1000 times and 100 electronic transitions could occur during one molecular vibration. The consequence is that during electron transfer, the nuclear positions change only insignificantly. This relationship is the Franck-Condon principle. For the electron transfer in our self-exchange processes, this means that it only takes place when the structure of the two reactants has been aligned by a distortion (or vibration). In Fig. 7.5, the potential energy of starting materials and products is given

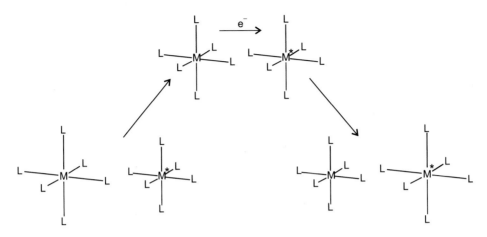

Fig. 7.6 Schematic representation of the electron transfer between two complexes. Before electron transfer can take place, the structures of the two reaction partners must align

as a function of the metal-ligand distance (with respect to one of the two reactants). Electron transfer can only occur at the intersection of the two potential curves. For the alignment of the structures, the energy difference U_i^* must be applied. In contrast to this, for an electron transfer starting from the equilibrium distance of the initial substance, a vertical transition with the energy λ_i must be provided, which is only possible e.g. through absorption of a photon. The energy required for this (internal reorganization energy of the complex) corresponds to the energy of the product at the equilibrium distance of the reactant and is four times the energy of the intersection of the two potential curves.

In Fig. 7.6, the electron transfer in a self-exchange process between two complexes is again schematically illustrated. The two starting materials have different metal-ligand bond lengths, as do the products. Before electron transfer can occur, the structures of the two reactants must align. At the transition state, the metal-ligand bond lengths are equal and lie between those of the oxidized and reduced species. This transition state corresponds to the intersection of the potential wells in Fig. 7.5. Now the relationship between the rate of electron transfer and the change in metal-ligand distance during electron transfer shown in Table 7.1 becomes clear. The smaller Δr is, the smaller is the energy U_i^* that has to be applied to reach the distorted equilibrium structure. Figure 7.5 schematically illustrates this relationship using iron(II/III) and ruthenium(II/III) as examples. In the case of ruthenium, Δr is significantly smaller. As a result, the two potential curves move closer together and U_i^* becomes smaller. Less energy is required to reach the distorted transition state.

How large the change in the M-L distance is depends on the occupation of the d orbitals and the ion radii in general. In general, large changes in the M-L distances are a consequence of the occupation of the antibonding e_g orbitals during the reduction step. A good example for this is the Co(III/II) pair. Cobalt(III) complexes usually exist in the low-spin state, where the three nonbonding orbitals are fully occupied (t_{2g}^6 configuration). The cobalt

(II) ion is usually present in the high-spin state and the antibonding orbitals are occupied by two electrons ($t_{2g}^5 e_g^2$ configuration). The resulting large change in distance when changing the oxidation state is accompanied by a very slow electron transfer. If only the number of electrons in the t_{2g} orbitals changes, as in the case of Ru(II/III), for example, the change in distance is small and rapid electron transfer is observed. Also in the Fe(II/III) example, only the occupation of the t_{2g} orbitals changes. However, here both iron centers are in the high-spin state and we consider a 3d element. The change in bond lengths is much more pronounced here than in the case of the 4d element ruthenium. If a rigid π acceptor ligand such as 2,2'-bipyridine is used, the changes in the bond lengths are compensated even further for ruthenium, which allows the electron transfer to take place even faster.

The Solvent Barrier   The third factor affecting the rate of electron transfer is the solvation barrier. In polar solvents, the metal ions are not present in isolation, but have other coordination spheres in addition to the first coordination sphere (the ligands), which are called the solvent shell. The strength of the interaction between a complex ion and the solvent shell depends on the charge of the complex ion. Higher charged complex ions have a more ordered solvent shell than lower charged complex ions. The ionic radius also plays a role. As both parameters change during the redox reaction, a rearrangement of the solvent shells occurs, creating an energy barrier. This process compares well with the steps discussed for the vibrational barrier. Similar to the internal structure, the solvent molecules (dipoles) must adopt a non-equilibrium orientation before electron transfer occurs. As the complex ion radius increases, the polarizing and thus ordering influence on the solvent molecules decreases and a faster reaction is favored. Non-polar solvents cannot form a solvent shell. Here, this barrier is completely absent.

Having considered the three rate-determining factors for self-exchange processes, we now turn to a redox reaction between two different complexes. This is then also referred to as a redox cross reaction. An example would be reaction c) in Fig. 7.4. Figure 7.7 shows the influence of the thermodynamic driving force of the reaction on the activation energy. We see that as the driving force ΔG^0 increases, the activation energy becomes smaller and smaller. If the difference in free energy between reactants and products is large enough, then the activation energy for the reaction disappears completely. However, this also means that if the energy difference ΔG^0 becomes even larger, the activation energy will increase again. This region is called the "inverted region" and some chemiluminescence reactions can be described with this model.

7.2.2   Inner Sphere Mechanism

The inner sphere mechanism assumes that a multinuclear complex is formed for electron transfer, whereby electron transfer is realized via a bridging ligand, as illustrated in Fig. 7.8. Such a situation is relevant, among other things, for electron transfer chains in biological systems.

Fig. 7.7 Schematic representation of the dependence of the activation energy on the thermodynamic driving force of the reaction ΔG^0

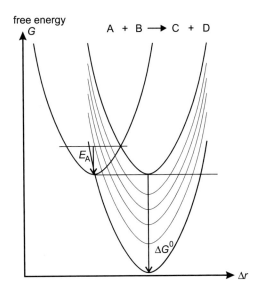

Fig. 7.8 Redox reaction according to the inner sphere mechanism. The electron transfer step takes place via a bridging ligand

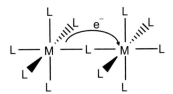

The inner sphere mechanism places high demands on the reactants. These include the presence of a potential bridging ligand in one of the complexes, the substitution of a ligand of the labile component and the formation of a bridging precursor complex. The mechanism consequently involves several individual steps, which makes a precise theoretical description difficult. Also, the experimental distinction between the inner sphere and outer sphere mechanism is not always clear.

An intramolecular electron transfer step in multinuclear complexes also proceeds according to the inner sphere mechanism. Such electron transfer reactions can be induced photochemically, for example, when the bridging ligand links two metal centers in different oxidation states. An example of this is the intense blue color of Berlin blue, which results from such an *Intervalence Charge Transfer (IVCT or IT)* transition between the iron(II) and iron(III) centers. In the Berlin blue (4 Fe^{3+} + 3 $[Fe(CN)_6]^{4-} \rightarrow Fe_4[Fe(CN)_6]_3$), an infinite $Fe(CN)_6$ lattice is present, with the CN bridging ligands coordinated to the iron(II) centers via carbon and to the iron(III) centers via nitrogen.

If we now recall our blue copper proteins, we find that nature got some things right (or the system was well optimized by evolution). Due to the protein environment, the intermolecular distance is fixed (formation of the outer sphere complex) and the vibrational barrier is extremely low because the complex is already in the distorted structure given by the protein environment (entatic state). Often such metalloproteins also have a very hydrophobic environment, so that the solvation barrier is also omitted. Thus, nothing stands in the way of a very fast electron transfer.

7.3 Non-Innocent Ligands Using the Example of NO

Redox reactions can occur not only between metal centers of complexes. When redox-active ligands coordinate to a redox-active metal centre, electron transfer can take place between ligand and metal centre. Because the oxidation state of the metal center and ligand is often difficult to determine for such complexes, these ligands are also referred to as "non-innocent" ligands. An example of such a ligand is nitric oxide (NO), whose complexes will be examined in more detail below.

NO is an important biomolecule that is involved as a messenger substance in a large number of physiological processes. NO complexes gained attention well before this realization. In the 1960s and 1970s, NO was used as an ESR-active probe to study the reaction of iron porphyrins (e.g., heme) and other biomolecules with small molecules (O_2, CO, ...). During this period (1974), a review article on transition metal-nitrosyl complexes by Enemark and Feltham appeared. [25] In this paper, the possibility of assigning (formal) oxidation numbers to the metal center and the nitrosyl ligand is discussed with the conclusion that the $M(NO)_x$ unit is best considered as a covalent unit. The next section will show that this approach does have its advantages. The Enemark-Feltham notation introduced in the article to classify transition metal-nitrosyl complexes has prevailed to this day. In 1992, NO was named Molecule of the Year, and in 1998, R. F. Furchgott, L. J. Ignarro, and F. Murad were awarded the Nobel Prize in Medicine for "Discoveries about the biomedical functions of nitric oxide."

7.3.1 Complexes with Redox-Active Ligands

The assignment of oxidation states for complexes with redox-active ligands (so-called non-innocent ligands) is often ambiguous and can lead to different results depending on the method used. For this reason, the Enemark-Feltham notation is still used to classify metal-nitrosyl complexes. In this method, the $\{M(NO)_x\}^n$-unit is considered as the covalent unit and, for further subdivision, the number of valence electrons of the metal center (usually corresponding to the d-electron number) together with the number of electrons in the π^* orbitals of the NO ligand is written as the number n at the curly bracket. In this way, the "correct" oxidation states of the metal center and nitrosyl ligand need not be determined, but any assumption can be made. At this point, we refer to the MO scheme of nitric oxide given in Fig. 4.13. However, if one wants to count the electrons in such compounds in order to estimate the stability of the complex according to the 18-electron rule, one can no longer avoid assigning oxidation states. For this case, IUPAC gives us some guidance. It specifies

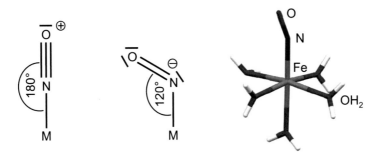

Fig. 7.9 Boundary cases of the M-N-O bond angle and structure of the complex $[Fe(H_2O)_5NO]^{2+}$.

that NO as a ligand is a neutral 3-electron donor. That is, the formal oxidation state is 0 and when counting electrons, three electrons are counted per NO ligand. In most cases, the formal oxidation state of the metal center and ligand determined in this way does not agree with the oxidation state determined spectroscopically or even with the aid of calculations. In the case of nitrosyl complexes, spectroscopic data such as the M-N-O bond angle, the N-O distance or the NO stretching vibration frequency provide possible clues. In most complexes, we assume that the nitric oxide is bound either as NO^+ (nitrosyl cation, isoelectronic to CO) or as NO^- (nitrosyl anion, isoelectronic to O_2). Figure 7.9 shows the theoretically expected M-N-O angle for both variants, which can be explained by the hybridization of the nitrogen atom. For the nitrosyl cation, there is a triple bond between the nitrogen and the oxygen. The nitrogen is sp hybridized and coordinates to the metal with a $180°$ angle (NO distance 1.06 Å, ν (NO) = 1950–1600 cm^{-1}). In contrast, the nitrosyl anion has a double bond and the nitrogen is sp^2 hybridized, so an angle of $120°$ is expected (NO distance 1.20 Å, ν (NO) = 1720–1520 cm^{-1}). The NO radical has a stretching vibrational frequency of 1906.5 cm^{-1} and a N-O distance of 1.14 Å.

In the further course, the relationship just described will be illustrated by three examples and, above all, also critically scrutinized. The first complex is the very stable compound [Fe(CO)$_3$NO]$^-$, with a linear {FeNO}^{10} unit and 18 valence electrons. This is followed by the very labile {FeNO}^{7} compound $[Fe(H_2O)_5NO]^{2+}$, which was also assumed to have a linear Fe-N-O arrangement for a long time until the first crystal structure of this complex was reported in 2019 [96]. The result can be seen in Fig. 7.9, the Fe-N-O angle is not linear. There are many theoretical considerations on this 19-valence electron complex in particular, which oscillate between the formulation iron(III)(HS)-NO$^-$ (S = 1), antiferromagnetically coupled, and iron(I)-NO$^+$. The last example is the nitroprussiate [Fe(CN)$_5$NO]$^{2-}$, a {FeNO}^{6} compound with 18 valence electrons, whose stability lies between the previous two.

The tetrahedral {FeNO}^{10} complex [Fe(CO)$_3$NO]$^-$ is formed by the reaction of [Fe(CO)$_5$] with nitrite in the presence of base (Ca(OH)$_2$). The reaction conditions alone show that the resulting complex is very stable (the aqueous solution is heated to boiling under reflux). The linear Fe-N-O angle (177°) indicates that the nitric oxide binds to the iron as NO$^+$. Somewhat contradictory at first appears the relatively long N-O bond (1.23 Å) and the very short Fe-N distance (1.63 Å). Both indicate a strong bond, which can be attributed to a pronounced π-backbonding. This can be excellently explained by the very low formal oxidation state of the iron (-2).

The $\{FeNO\}^7$ complex $[Fe(H_2O)_5NO]^{2+}$ is responsible for the brown color in nitrate detection (ring test).

$$4\,Fe^{2+} + NO_3^- + 4H^+ + 3\,H_2O \rightarrow \left[Fe(H_2O)_5NO\right]^{2+} + 3\,Fe^{3+}$$

This compound is very labile and could only be isolated as a solid in 2019. The Fe-N-O angle is already clearly angled at $160°$, but still far from $120°$. The unstable character of this compound can be well explained by the 18-electron rule. Calculations suggest that this compound is most likely an iron(III) complex with coordinated NO^-. [26] Also discussed is the possibility iron(II)-NO(radical), where a linear Fe-N-O angle is nevertheless possible. Ultimately, a very recent study has shown that all three possibilities, iron(III)-NO^-, iron (II)-NO, and iron(I)-NO^+ are relevant moieties within this covalent unit to discuss the chemical bond situation and thus none can be classified as false [96]. Also, the use of the M-N-O angle to assign charges or oxidation states rather rarely leads to a correct result [97].

In the nitroprussiate $[Fe(CN)_5NO]^{2-}$, an octahedral complex, the N-O distance is significantly shorter (1.13 Å) and the Fe-N distance is significantly longer (1.67 Å) compared to the $\{FeNO\}^{10}$ complex. Both complexes are 18-electron complexes with a linearly bound NO^+ (Fe-N-O angle at $[Fe(CN)_5NO]^{2-}$ $176°$). The differences in bond lengths are due to a weaker π backdonation in the nitroprussiate. In this compound, the iron is present in the formal oxidation state $+2$ and thus fewer d electrons are available to it for π backbonding. This results in a weaker binding of the NO^+ to the iron centre. This is exploited, for example, in the use of sodium nitroprussiate as a drug (under physiological conditions, the biomolecule NO is split off, which is responsible for the regulation of blood pressure, among other things).

The three examples show us that even the spectroscopic data do not always allow a doubtless or meaningful assignment of the oxidation states. To explain this relationship, we must resort to molecular orbital theory; all other models fail. A strongly covalent multiple bond is formed between the metal and the ligand. The electrons are delocalized in molecular orbitals via the Fe-NO unit. As a result, the whereabouts cannot be unambiguously determined, nor can formal oxidation numbers (which are anyway only computational aids) be determined. In Fig. 7.10, a section of the molecular orbital (MO) scheme of an iron nitrosyl complex is given. In the initial complex, the iron(II) ion is surrounded by a macrocyclic chelating ligand in a square planar fashion. Note the strong destabilization of the $d_{x^2-y^2}$ orbital! The three d orbitals with z component interact with the two π^* orbitals of NO and five molecular orbitals are formed, the structure of which are given in Fig. 7.10 on the right. The molecular structure, also shown, indicates that an angled Fe-NO unit is obtained in this $\{FeNO\}^7$ complex. The Fe-N-O angle is $140°$, the N-O distance is 1.19 Å, and the Fe-N distance is 1.73 Å. These data correspond most closely to an iron(III) complex to which an NO^- is attached. During complex formation, an intramolecular electron transfer from the iron to the NO has occurred.

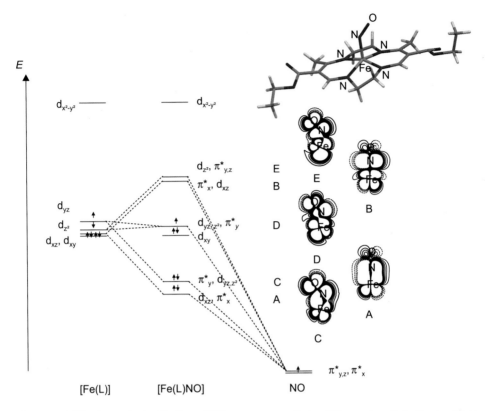

Fig. 7.10 MO scheme for the binding of NO to a macrocyclic iron(II) complex. Only the d-orbitals of iron and the π^* orbitals of NO were considered

7.4   Questions

1. Draw the energetic splitting of the d-orbitals for a stretched and compressed octahedron starting from the ideal octahedron!

2. Discuss whether a regular octahedron can exist as a coordination polyhedron in $[CoL_6]^{2+}$ complexes.

3. For which d^n configurations of octahedral complexes do you expect a Jahn-Teller distortion? If necessary, distinguish between the high-spin and the low-spin case.

4. Explain the different changes in bond lengths of the redox pairs $[Co(NH_3)_6]^{2+/3+}$, $[Fe(H_2O)_6]^{2+/3+}$ and $[Ru(NH_3)_6]^{2+/3+}$ considering the electron configuration. Why does the electron transfer of $[Ru(bipy)_3]^{2+/3+}$ appear to proceed with almost no structural change?

5. For the NO complexes presented in this chapter, determine the electron number using the 18-electron rule and Enemark and Feltham notation.

Supramolecular Coordination Chemistry

8

While molecular chemistry is the chemistry of covalent bonds, supramolecular chemistry deals with chemistry beyond the molecule. It refers to organized complex units that are held together by intermolecular interactions (hydrogen bonds, van der Waals forces). These supramolecular units are called supermolecules. This term was introduced as early as the mid-1930s to describe more highly organized units resulting from the assembly of coordinatively saturated species. Supramolecular chemistry unfolds mainly in the frontier areas dealing with physical and biological phenomena at the molecular level. New physical properties lead to new material properties. A good example for this, which is already widely used in applications, are liquid crystalline systems. The host-guest relationships that frequently occur in biological systems or the lock-and-key principle are also phenomena that can be explained with the help of supramolecular chemistry. Molecular machines (sometimes called nanomachines) are supramolecular entities composed of macromolecules that can perform specific functions or movements. For the design and synthesis of molecular machines, Jean-Pierre Sauvage, Sir J. Fraser Stoddart, and Bernard L. Feringa were awarded the Nobel Prize in Chemistry in 2016.

> **Definition of Supramolecular Chemistry According to Lehn**
> "Supramolecular chemistry deals with structures and functions of units formed by association of two or more chemical species."

8.1   Molecular Recognition

The first research projects that led to the development of supramolecular chemistry dealt with the recognition of spherical substrates and the formation of cryptands. For the work carried out in this regard, Charles J. Pedersen (discovery of crown ethers), Donald J. Cram (molecular host-guest relationships) and Jean-Marie Lehn (helicates, cryptands) were awarded the Nobel Prize in Chemistry in 1987 [27–29]. These projects were motivated by biochemical questions (selective complexation of alkali metal cations) or questions regarding anion activation. In Fig. 8.1, three crown ethers and two macrobicyclic cryptands are shown with their respective names. The crown ethers belong to the macrocyclic ligands of the coronand type. When naming a crown ether, the number of atoms in the macrocyclic polyether is given first, followed by the number of donor atoms. The unique property of these ligands is the ability to coordinate alkali metal ions that are normally not bound by other ligands. Particularly good properties as complexing agents are achieved when the oxygen atoms are always separated by two carbon atoms. Then chelate five-membered rings are formed with the metal center, which we already know to be particularly stable. The stability of the corresponding complexes can be explained by the macrocyclic effect and the HSAB concept (the hard oxygen serves as a donor atom for the alkali and earth alkali metal cations, which are also hard). As can be seen in Fig. 8.1, the crown ethers have a specific hole size. This is used to denote the cavity available for complexation of a cation in a given crown ether. The size range is 1.2–1.5 Å for [12] crown-4, 1.7–2.2 Å for [15] crown-5 and 2.6–3.2 Å for [18] crown-6. The alkali metal cations also have different sizes with 1.36 Å for Li^+, 1.94 Å for Na^+, and 2.66 Å for K^+. Depending on the hole:cation size ratio, 1:1 complexes can be selectively obtained with the respective cation as given in Fig. 8.1. In this case, the cation that best fills the hole is coordinated. By complexing the cation with the crown ether, otherwise only water-soluble alkali metal salts can be dissolved in organic solvents (e.g. dichloromethane). This allows the anions to be activated for subsequent reactions.

A further increase in binding selectivity can be achieved if 3-dimensional ligand cages, known as cryptands, are used. Two examples of this are given in Fig. 8.1 below. Again, the size of the cavity can be varied by the length of the bridges. The cryptands are also well suited for the complexation of alkali and earth alkali metal cations. Compared to the macrocyclic ligand, the stability of the corresponding complexes is further increased. In this context, one then speaks, in reference to the macrocyclic effect, of the cryptand effect.

Further development of these areas of work led to questions such as the targeted recognition of anionic substrates, tetrahedral substrates, ammonium ions and related substrates, as well as multiple recognition (bi- and polynuclear cryptands, linear substrates). The work on coronands and cryptands led to further areas of research, three examples of which are given here.

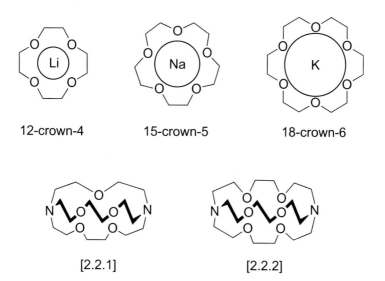

12-crown-4   15-crown-5   18-crown-6

[2.2.1]   [2.2.2]

Fig. 8.1 Top: Structure of three crown ethers with an illustration of the different hole sizes responsible for the binding selectivity for the cations Li^+, Na^+, and K^+. Bottom: structure of two macrobicyclic cryptands. The donor sites per bridge are counted in the given abbreviated designation

- Supramolecular reactivity and catalysis
- Transport processes (e.g. ion channels in biological systems) – Carrier and channel design
- Molecular and supramolecular functional units – Chemionics, molecular machines

Molecular functional units are defined as structurally organized and functionally connected chemical systems that are incorporated into supramolecular structures. On the one hand, these can be the supermolecules already defined, i.e. a functional unit consisting of receptor and substrate. Another subfield are molecular aggregates. These are defined as polynuclear systems formed by spontaneous association of a larger number of components. Of particular interest to the coordination chemist is the field of systems with the ability to self-assemble. A very well-known example is the spontaneous formation of the double helix of nucleic acids, which is determined by the given pattern of base-base interactions. Molecular recognition and positive cooperativity are the main characteristics of this mechanism. The previous examples were mainly about pre-organized systems (receptors for recognition, catalysis, and transport processes). The next stage is self-organizing systems that spontaneously self-assemble from their components under certain conditions. These systems can therefore be called programmed molecular and supramolecular systems that generate ordered units according to a defined plan based on molecular recognition. In this context, we turn to what is probably the best-known example – the self-assembly of metal complexes with a double-helix structure, the helicates. Before doing so, we will deal with bond selectivity and molecular recognition in more detail. The molecular recognition already introduced in the crown ether example can also be used to selectively form bonds. This idea is exploited in the template synthesis of macrocyclic ligands.

8.1.1 The Template Effect

An important principle of supramolecular chemistry is bond selectivity, thus molecular recognition. This selectivity has a crucial importance in template syntheses of macrocyclic ligands. Macrocyclic ligands are of particular interest to the coordination chemist. On the one hand, the complexes are characterized by a particular stability and, as we have already seen with crown ethers, binding selectivity. In addition, in biological systems, the active site of metalloenzymes is often complexed by a macrocyclic ligand. To achieve a better understanding of these enzymes, there is interest in preparing model compounds with macrocyclic ligands. The synthesis of cyclic organic compounds is not straightforward as a variety of by-products can be obtained. In order to achieve specific reaction control, one can exploit the template effect. In our case, the template is a metal ion. The information stored in the metal ion leads to selectivity (of a certain reaction), which is stored in the (resulting) ligand structure.

Definition of Template Synthesis According to Busch
A chemical template organizes a collection of atoms with respect to one or more geometric location(s) to achieve a particular interconnectedness of atoms.

In general, the template serves to pre-orient the components to be linked, so that the formation of undesired by-products is prevented. Another alternative for the targeted synthesis of, for example, macrocyclic compounds would be to dilute the components to a high degree in order to suppress oligomerization or polymerization this way. A common problem in template synthesis is the removal of the metal ion to present the free ligand. The key to successful template synthesis is the selection of a suitable cation. To do this, we must first ask what kind of molecular information can be stored in a cation. The properties of cations are mainly determined by their ionic radius and valence electron configuration. From this, the following molecular information can be derived.

- preferential coordination number
- preferred coordination geometry (e.g. square planar for Cu^{2+} and tetrahedral for Cu^+)
- preferred donor atoms (HSAB concept)
- Ion radius (e.g. crown ether)

Therefore, alkali metal ions are particularly suitable for the preparation of crown ethers, while soft cations are preferred for ligands with N and S donor atoms. The ionic radius plays an important role in determining the size of the macrocycle. An example of this is given in Fig. 8.2. Similar criteria play a role in the synthesis of cage compounds. Molecular information may also be stored in the ligand. A good example of this is given in Fig. 8.3.

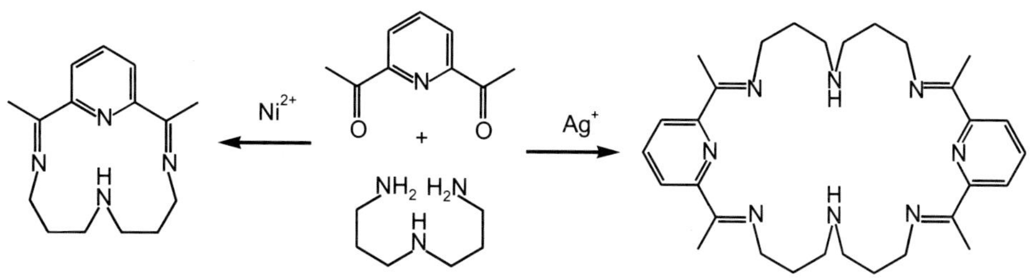

Fig. 8.2   Influence of the ionic radius of the template on the size of the macrocycle formed

- preferred donor atoms (HSAB)
- preferred coordination geometry (e.g. *fac* and *mer*)
- Flexibility

This allows us to redefine the molecular recognition already illustrated for crown ethers.

Definition of Molecular Recognition

Molecular recognition is a matter of storing and reading out information at the supramolecular level. The molecular information is encoded in the structure of the ligand (Fig. 8.3) and the transition metal ion (Fig. 8.2). The metal ion has preferred coordination numbers and geometries, and the ligand leads to specific geometries through fixed steric demands.

Templates can be divided into two different categories, the kinetic template and the equilibrium or thermodynamic template. Figure 8.4 shows the basic principle of a thermodynamic template. The thermodynamic or equilibrium template favors the formation of a product that is in equilibrium with many other products. The template removes the desired

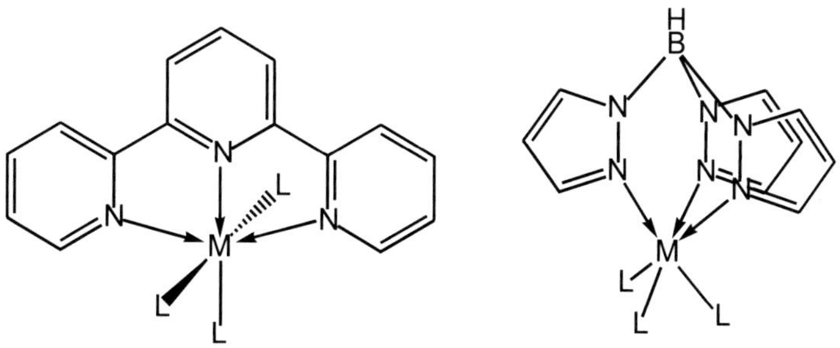

Fig. 8.3   Molecular information can be stored in the ligand (*mer* coordination for terpyridine and *fac* coordination for trispyrazolyl borate in octahedral complexes) and in the metal ion (coordination number, coordination geometry, preferred donor atoms)

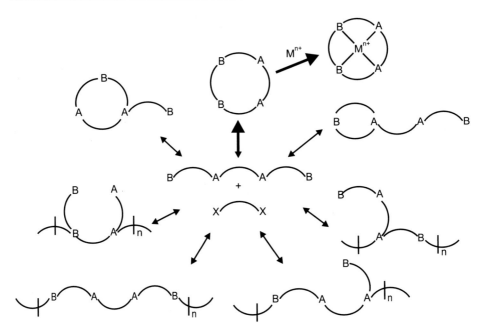

Fig. 8.4 Basic principle of an equilibrium or thermodynamic template. The desired product (macrocycle) is removed from the equilibrium by the template (metal ion) and the yield is thus increased

product from equilibrium. Following le Chatelier's principle, the desired macrocycle is replicated in Fig. 8.4 and the yield is increased.

The kinetic template, given in Fig. 8.5, influences the reaction sequence to arrive at the desired product. In contrast to the equilibrium template, where all reactants are converted in one pot, several steps are carried out. This ensures that the desired final product is obtained.

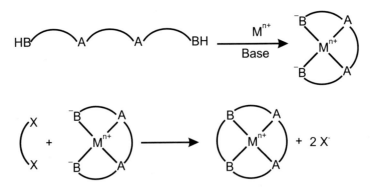

Fig. 8.5 Basic principle of a kinetic template. The macrocycle synthesis is subdivided into several individual steps

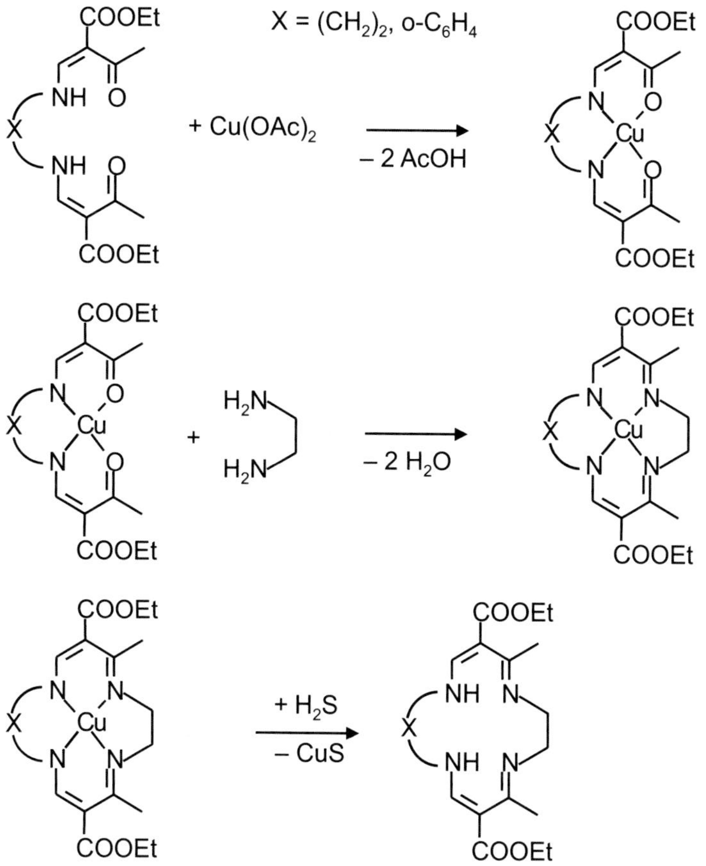

Fig. 8.6 Example of the synthesis of a macrocyclic complex (kinetic template) with subsequent removal of the template to generate the free ligand [20]

A critical step in template synthesis is often the subsequent removal of the template from the formed macrocycle to obtain the free ligand. Figure 8.6 shows an example of the synthesis of a macrocyclic ligand. The copper(II) ion, which prefers a square planar coordination environment and is therefore well suited for the synthesis of planar N_4 macrocycles, is used as a template. In the final step, the copper is removed from the complex formed. The low solubility of copper sulfide is used for this purpose.

8.2 Helicates

The concept of molecular recognition used in template synthesis can be used to build larger and more complex structures, leading to supramolecular coordination chemistry. A prerequisite for the assembly of supramolecular structures is that the compounds are at the same

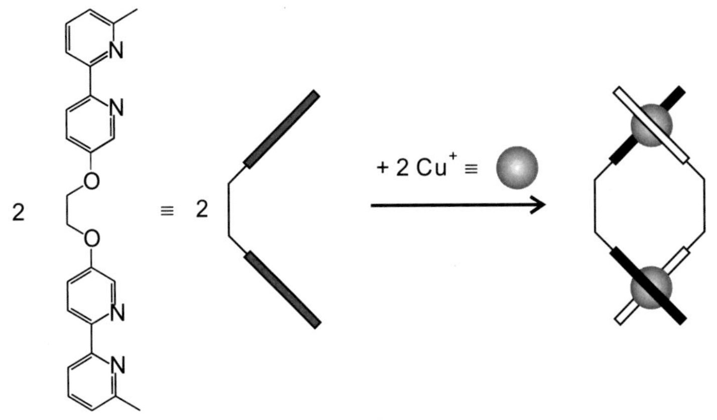

Fig. 8.7 Binuclear copper "helicate" complex

time kinetically labile and thermodynamically stable. As for the thermodynamic template, several possible reaction products are in equilibrium. Sufficiently long reaction times lead to the thermodynamically most stable product being the main product. If these boundary conditions are fulfilled, supramolecular coordination chemistry can be carried out. The basic principle of this chemistry can be well illustrated by Lehn's "helicates". As an example, the reaction of the copper(I) ion with 2,2'-bipyridine derivatives is given in Fig. 8.7. Copper(I) ions, as d^{10} systems, prefer a tetrahedral coordination environment. This implies that two 2,2'-bipyridine units always coordinate to one copper ion. Since the bridge between the two 2,2'-bipyridine units is too short, monomeric complexes cannot be formed. Conceivably, dimeric (helical), trimeric (triangular) or tetrameric (square) complexes are formed instead. Characterization of the reaction product reveals that in the example shown in Fig. 8.7, only one variant is specifically obtained, the helix. This is the thermodynamically most stable structure, which is obtained as the end product if the reaction time is sufficiently long.

In the second example shown in Fig. 8.8, in addition to the two products shown and the possibilities already mentioned (triangle, square, ...), mixed helicates with a triple and a quadruple complexing ligand are also conceivable. In this case, either part of the octadentate ligand would not be bound to a copper(I) ion, or the copper(I) ion coordinated to it would lack two ligands to saturate its coordination environment. For this reason (again, given a sufficiently long reaction time), the mixed, thermodynamically unfavorable variant is not observed, although it is statistically preferred. Lehn's helicates attracted attention not only because of the basic ideas of supramolecular chemistry that can be illustrated very well by these examples. They are the first synthetic compounds with helical structure, which are a model for the double helix of DNA.

For the formation of helicates, the bridge between the 2,2'-bipyridine units in the ligand must be very flexible. What happens when the bridge is shortened and thus becomes more rigid is shown in Fig. 8.9. In this example, the metal center used is the iron(II) ion, which

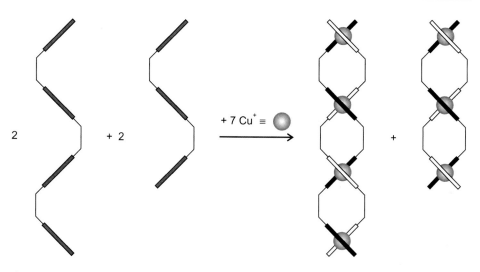

Fig. 8.8 Self-organisation using the example of trinuclear and tetranuclear copper "helicate" complexes

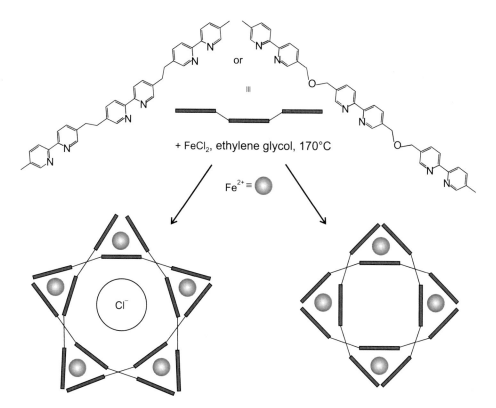

Fig. 8.9 Influence of the flexibility of the ligand on the supramolecular aggregates formed [30]

prefers an octahedral coordination environment. Therefore, three 2,2'-bipyridine units coordinate to the ferrous ion in this example. Star-shaped structures are obtained as products, the size of which depends on the flexibility of the ligand. The more flexible the ligand, the smaller the polynuclear complex. At this point a brief comment on the reaction conditions suggests itself. As indicated in Fig. 8.9, comparatively high reaction temperatures are required for the coordination chemistry, which led to the choice of the solvent used (ethylene glycol = 1,2-ethanediol, boiling point 197 °C). High temperatures and very good solubility of the starting materials and intermediates are necessary to ensure that the also conceivable by-products remain in solution and are not removed from the equilibrium.

The examples shown (there are countless more) can be summarized in the following definition of self-organization.

Definition of Self-Organisation
The simple self-assembly of molecular building blocks requires binding forces, while self-organization also requires information. Self-organizing systems spontaneously assemble themselves from their components under certain conditions.

8.3   MOFs - *Metal-Organic* Frameworks

For the examples discussed so far, discrete supramolecular units have been obtained. The next step, the construction of one-dimensional polymers, or 2- or 3-dimensional networks, leads us to MOFs. In 2009, an IUPAC working group was established to find guidelines for the definition and distinction of the terms MOF (= metal-organic framework) and coordination polymer (CP). The problem lies in the interdisciplinary nature of this area of research, which combines inorganic chemistry, solid state chemistry, and coordination chemistry. Accordingly, there are research groups that use the terms MOF and 3-dimensional coordination polymer as synonyms, while for others MOFs are porous, crystalline framework compounds whose nodes consist of multiple metal clusters. The IUPAC recommendations published in 2013 are presented and applied within this book [31].

8.3.1   Coordination Polymer or MOF?

The term coordination polymer refers to coordination compounds consisting of repeating building units that extend in 1, 2, or 3 dimensions. Figure 8.10 gives an example of a monomeric complex (only one metal ion) and an IUPAC-compliant 1-dimensional coordination polymer derived from it. Coordination compounds with repeating units in two or

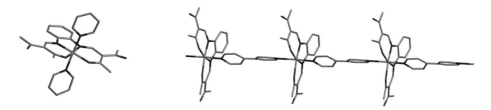

Fig. 8.10 Transition from a monomeric complex *(left) to* a 1D coordination polymer *(right)*

three directions can be referred to as 2D or 3D coordination polymers, or as coordination networks. In the context of this book, we will use the term coordination network. This term can also be used to refer to linked strands of 1D coordination polymers. A property often associated with MOFs is porosity, and IUPAC recommends that this term is used for coordination networks with organic ligands that have potential voids. The voids may be filled, or accessible only under certain conditions. The examples shown in this book are 2- and 3-dimensional MOFs whose nodes are composed of metal clusters held together by organic ligands called linkers, and are porous and crystalline. In reference to purely inorganic porous 3D structures, the zeolites, the nodal points are also called SBU = secondary building unit. According to IUPAC, crystallinity as well as the presence of metal clusters as nodes is not a necessary prerequisite for calling a coordination network a MOF.

8.3.2 The Structure of MOFs

MOFs are porous hybrid materials, formed in a reaction between organic and inorganic species to build a three-dimensional framework structure. Most MOFs are produced by classical solid-state chemical methods such as solvothermal synthesis. In this process, the starting materials are reacted in a suitable solvent at high temperatures and pressures. Under these conditions, among other things, the solubility of the individual components is improved and by-products are avoided.

The skeleton of the metal-organic framework compounds contains both inorganic (nodes, SBU) and organic units (the "linkers"), which are held together by strong bonds, usually coordinative bonds. Conceptually, there is no difference between the classical inorganic porous solids and the inorganic/organic hybrid compounds. In both cases, the three-dimensional skeletons are obtained by the assembly of so-called secondary building units (SBUs). In Fig. 8.11, two examples of the inorganic nodes, SBUs, are given. These are multinuclear complexes in which the ligands (acetate and water in the examples shown) can be replaced by bridging ligands. The SBUs can be equated to geometric solids. For example, the basic zinc acetate forms a molecular octahedron, while the copper acetate is a molecular square if only the acetate ligands are considered. The additional exchange of the water ligands also leads to a molecular octahedron.

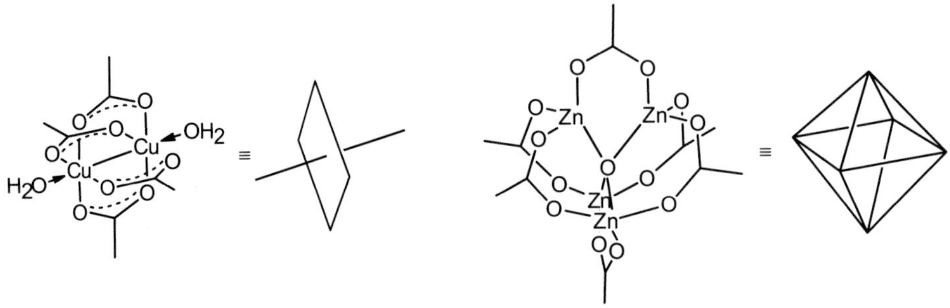

Fig. 8.11 Various secondary building units (SBUs). Shown on the left is copper acetate, a dimeric complex with the general composition $[M_2(O_2CR)_4L_2]$. If only the acetate ligands are considered as attachment points for the organic linkers, a molecular square is obtained that can be used to construct two-dimensional layers. The basic zinc acetate shown on the right with the general composition $[M_4O(O_2CR)_6]$ forms a molecular octahedron

Bridging organic ligands are used as linkers in MOFs. This leads to the fact that the framework, in contrast to the purely inorganic zeolites, is now primarily composed of covalent bonds. Rigid ligands with multiple bonds must be used to ensure the desired porosity. Figure 8.12 shows various organic linkers. These are usually rigid molecules with two to four functional groups for network formation. The distance between the nodes can be regulated via the organic framework. In Fig. 8.12, only a small excerpt of the possible bridging ligands that can be used for the synthesis of MOFs is given. The combination with different SBUs provides a wide range of MOFs for which the properties (e.g. the pore size) can be systematically varied.

Parallels to the classical, purely inorganic, porous materials can also be seen in the naming. As for zeolites, new compounds are named by three letters (usually the geographical origin of the new compound) and a number. For example, *MIL* stands for materials from the Lavoisier Institute. The very first representatives were designated MOF and a consecutive number.

As a first example, we consider MOF 5 with composition $Zn_4O(BDC)_3$. In this MOF, the SBU is the basic zinc acetate shown in Fig. 8.11, which forms a molecular octahedron. When the acetate ions are replaced by the bridging ligand (linker) terephthalate, we obtain a porous three-dimensional framework structure as shown in Fig. 8.13. MOF 5 was prepared by Yaghi in 1999 and was the first simple coordination network with a very high specific surface area (2900 m^2 g^{-1}). After fabrication, the pores are initially filled with solvent, which is relatively easy to remove. Into the now empty pores, 1.04 cm^3 g^{-1} nitrogen can be condensed, which is comparable to the adsorption capacities of activated carbon. In this MOF, the pore size can be nicely controlled by varying the linker. By using naphthalene dicarboxylate or pyrendicarboxylate anions (IRMOF-8 and 14), the pores are systematically enlarged. All examples have the same cubic network, which is why these MOFs are called IRMOFs, "Isoreticular Metal-Organic Frameworks" [32, 33].

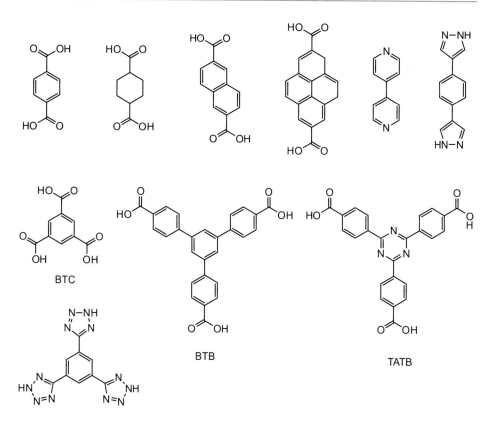

Fig. 8.12 Examples of bridging carboxylate-based or heterocyclic-based ligands. The ligands can have two, three or (not shown here) four functional groups to form a network. The spacing between the functional groups can also be controlled. As an example, terephthalic acid (benzenedicarboxylic acid *BDC*), cyclohexanedicarboxylic acid, naphthalenedicarboxylic acid, and pyrenedicarboxylic acid are given at the top from left. The same is possible for the ligands with three functional groups (benzene tricarboxylic acid or trimesic acid *BTC* and benzene tribenzoate *BTB*). The ligands can be further functionalized by introducing heteroatoms (4,4',4''-s-triazine-2,4,6-triyltribenzoate *TATB*) to achieve new or additional properties

As a second example, we build a MOF starting from copper acetate. First, we replace the acetate ions with *trans-1*,4-cyclohexanedicarboxylate. In this process, a two-dimensional layer of cross-linked copper centers is obtained. In a second step, the water is replaced by the linker 4,4'-bipyridine (bipy) and we again arrive at a porous 3D framework. In contrast to MOF 5 and the IRMOFs, this network is no longer cubic but composed of cuboids. Also in this example, the pore size can be varied by using different dicarboxylic acids. Instead of 4,4'-bipyridine, other bipyridines can be used as linkers. In the example shown in Fig. 8.14, interpenetrating networks are observed in the crystal structure. This significantly reduces the effective pore size – a not necessarily desirable effect [34].

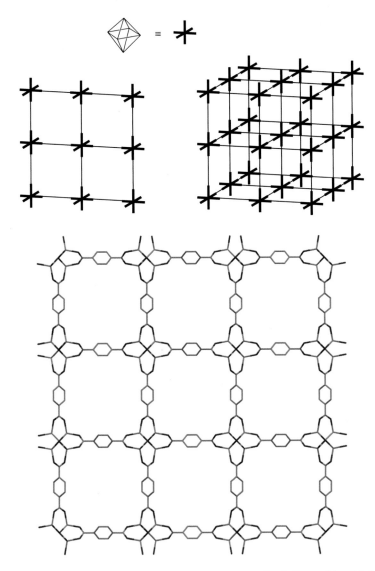

Fig. 8.13 *Top:* schematic structure of MOF 5 ($Zn_4O(BDC)_3$ with BDC = 1,4-benzenedicarboxylate). A porous cubic network is obtained. *Bottom:* Detail of the crystal structure

8.3.3   Advantages and Potential Applications

The first advantages of MOFs over zeolites can already be found in the synthesis. While the latter require inorganic or organic templates to achieve the desired pores, which are costly to remove, in MOFs the solvent is often also the template. As a result, the skeleton of MOFs is usually neutral and stable even after the template is removed. Through the interaction of inorganic and organic parts, hydrophilic and hydrophobic regions can be realized.

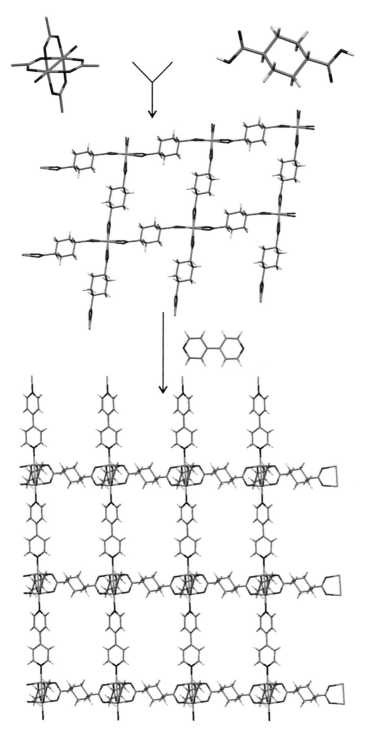

Fig. 8.14 Structure of a MOF starting from copper acetate with *trans-1*,4-cyclohexanedicarboxylate and 4,4′-biypridine as linker. The exchange of the acetate ligands with the bridging *trans-1*,4-

Furthermore, MOFs exhibit a much greater variability. This starts with the possible cations for the construction of the network (almost all bivalent to tetravalent metal ions) and increases again significantly with the organic bridging ligands, which can be additionally functionalized. A major disadvantage of MOFs compared to the purely inorganic zeolites is the significantly lower thermal stability and the lower stability towards acids and bases. The synthesis of MOFs is also not as simple as it appears on paper, as a large number of undesirable by-products are possible.

The diversity and variability of MOFs result in an equally diverse range of applications, a few of which are given below [35].

- Gas storage (hydrogen, methane)
- Gas cleaning, gas separation
- Sensors
- heterogeneous catalysis

All applications are based on the large surface area of MOFs, their porosity. The high application potential of MOFs lies in the possibility to selectively tailor the shape and size of the pores. Here, MOFs show a higher flexibility than classical porous solids such as zeolites.

Gas Storage The possibility of storing hydrogen is being investigated very intensively. Here, vessels filled with MOFs show a significantly higher uptake capacity than empty vessels at low pressures. This observation is attributed to adsorption effects. For example, studies on MOF-5 showed that 1.3 wt% H_2 can be adsorbed at 77 K (1 bar) [36]. Further investigations have shown that as many small pores as possible (just large enough for H_2) and free coordination sites at the metal have a beneficial effect on hydrogen storage. Both were realized in the following example (again with copper acetate). A 3D network with free coordination sites on the metal can be obtained starting from copper acetate if tricarboxylates are used instead of linear bridging dicarboxylates, as shown in the example in Fig. 8.15. The many small pores are realised by two interpenetrating networks and the free coordination sites at the metal are formed when the water on the copper is removed. Gas sorption experiments confirm the very high porosity of the compound. For hydrogen storage it can absorb about 1.9% hydrogen (10.6 mg/cm^3) under normal pressure at 77 K, which is (until 2006) one of the highest values obtained for MOFs [37].

Different approaches are used in heterogeneous catalysis. In a classical variant, (a) metal nanoparticles are deposited on the surfaces in the pores of the MOFs (metal@MOF), this

---

Fig. 8.14 (continued) cyclohexanedicarboxylate leads to the formation of a 2D network. The additional exchange of the coordinated water with the bridging 4,4'-bipyridine leads to the formation of a rigid 3D network. Interpenetrating networks are observed in the crystal structure

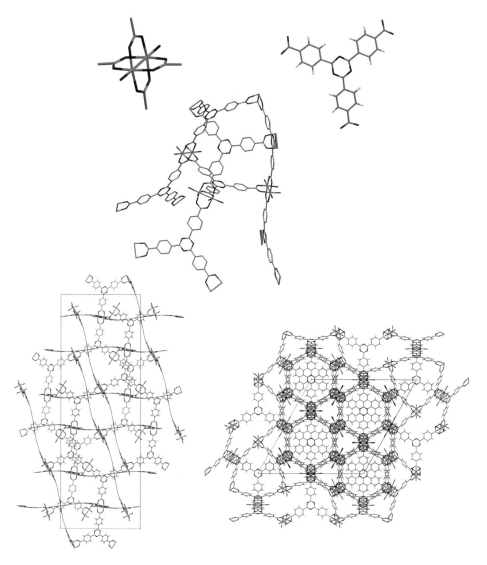

Fig. 8.15 Structure of a MOF starting from copper acetate with free/flexible coordination sites on the metal: $Cu_3(TATB)_2(H_2O)_3$ with TATB = 4,4′,4″-s-triazine-2,4,6-triyltribenzoate. The tridentate ligand gives rise to a three-dimensional framework compound. The water attached to the copper in the figure can be removed following synthesis

approach is already known from zeolites. Other alternatives are (b) the linker molecules as active sites, (c) the SBUs as active sites or (d) the introduction of active sites by subsequent modifications.

8.4   Questions

1. What is molecular information and where can it be stored in?
2. Why is not water but methanol used as solvent for the synthesis of the helicates?
3. Why have the copper sites in helicates a tetrahedral and not a square planar coordination geometry?
4. Are other products conceivable for the reaction in Fig. 8.7?
5. Explain why no mixed complexes are formed in Fig. 8.8?
6. What is a MOF?
7. What advantage do MOFs have over alternative porous materials? What are the disadvantages?

Metal-Metal Bond

9

In the last section, we considered a whole series of multinuclear complexes. They all had in common that they were linked by bridging ligands. In this section we turn to compounds in which a covalent bond is discussed between two or more metal atoms. In introducing the 18-electrons rule (Sect. 3.2), the possibility of metal-metal bonds has already been mentioned. The formulation of such bonds deviates greatly from Werner's ideas about the molecular chemistry of transition metals and the breakthrough did not occur until the 1950s, when X-ray structural analysis clearly demonstrated the presence of metal-metal bonds.

Clusters, or more precisely cluster complexes (since the term cluster can also carry another meaning in chemistry and physics) are, according to IUPAC, complexes with three or more metal centers with metal-metal bonds [12]. Colloquially, the dinuclear systems are often included and also in this chapter, dinuclear systems are used as first examples in the following. Cluster complexes can be divided into two classes: One is halido complexes of the d electron-poor transition metals, which also include analogous oxido and thiooxido complexes. The combination of d electron-poor metal centers and π donor ligands leads to stable systems with metal-metal bonding. These compounds often do not obey the 18-electron rule and can be better described by ligand field theory due to their ionic structure. The second class of compounds are polynuclear carbonyl complexes of d electron-rich transition metals with π acceptor ligands. For these complexes, which are usually covalently structured, the 18-electron rule can be applied to determine the number of metal-metal bonds. Each metal-metal bond contributes one electron per metal center to the electron balance.

© The Author(s), under exclusive license to Springer-Verlag GmbH, DE, part of
Springer Nature 2023
B. Weber, *Coordination Chemistry*,
https://doi.org/10.1007/978-3-662-66441-4_9

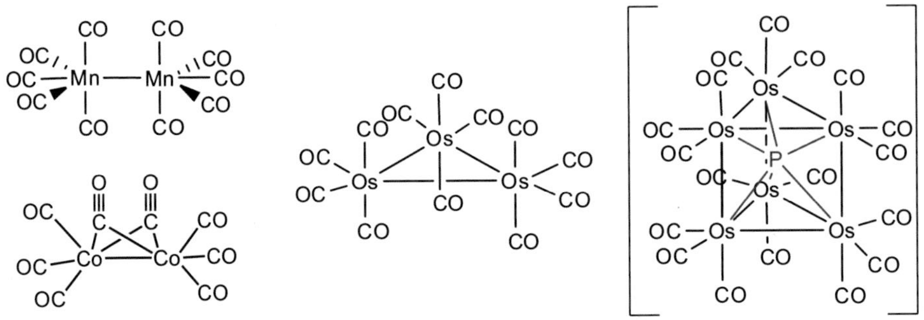

Fig. 9.1 Multinuclear carbonyl complexes with metal-metal bonding

9.1   Nomenclature for Polynuclear Complexes/Metal-Metal Bonds

For multinuclear complexes, a metal-metal bond is indicated at the end of the name. In round brackets, the two metal centres between which a metal-metal bond exists are indicated joined by a hyphen. As an example, we name the dinuclear cobaltcarbonyl complex in Fig. 9.1 It is called di-μ -carbonylbis[tricarbonylcobalt(0)] *(Co-Co)*. The dinuclear manganese complex with composition $[Mn_2 (CO)_{10}]$ is the decacarbonyldimanganese *(Mn-Mn)* or the bis(pentacarbonylmanganese) *(Mn-Mn)*.

9.2   Metal-Metal Single Bond

9.2.1   The EAN Rule

For some metal carbonyls, the number of metal-metal bonds can be determined via the 18-electron rule. For this purpose, the EAN rule (Effective Atomic Number Rule) derived from this rule can be used. This can be considered as an extension of the 18-electron rule for clusters. As for the 18-electron rule, the system strives to achieve a noble gas configuration. The "missing" valence electrons can be filled in by forming metal-metal bonds, since each metal-metal bond contributes one electron per metal center to the electron balance. Given the complex formula, with the corresponding total valence electron number N (no metal-metal bonds considered) and the number of metal centers n, the number of metal-metal bonds x can be predicted or calculated as follows:

$$x = \frac{18n - N}{2}$$

Our first example is decacarbonyldimanganese [Mn_2CO_{10}]. If we assume that there are two monomeric complexes of the composition [$Mn(CO)_5$] without metal-metal bond, then the number of valence electrons gives us 17; 7 from the manganese and 10 from the five carbonyl ligands. As a monomer, the carbonyl complex is not stable. If a metal-metal bond is now formed, then one more electron per manganese is contributed to the electron balance and we obtain 18 valence electrons per metal center or 36 for the whole complex. With metal-metal bond, the complex is stable. If we want to apply the EAN rule, then $n = 2$; $N = 34$ and consequently $x = 1$. The complex has one manganese-manganese bond. Our second example is the dinuclear cobalt complex with the two bridging carbonyl ligands given in Fig. 9.1. If we assume that there is no metal-metal bond, then we again get 17 valence electrons per cobalt; 9 from the cobalt, 6 from the three terminal carbonyl ligands, and two from the two μ_2-bridging carbonyl ligands. Caution. The carbonyl ligand can always contribute a maximum of two valence electrons, regardless of the bridging mode. The two cobalt atoms must share the free electron pair from the bridging carbonyl ligands! By forming a metal-metal bond, the noble gas configuration is achieved. In the EAN rule, $n = 2$; $N = 34$ and consequently $x = 1$. As further examples, let us look at two slightly larger clusters. For the trinuclear complex [$Os_3(CO)_{12}$], $n = 3$, $N = 48$, and $x = 3$. The three metal-metal bonds can be realized if the three osmium atoms form a triangle. For the complex [$Os_6(CO)_{18}P$]$^-$, we can assume for electron counting that the osmium and the carbonyl ligands are uncharged and the phosphorus atom, which is then simply negatively charged, has 6 valence electrons as a ligand. This gives us $n = 6$, $N = 90$, and $x = 9$. Nine metal-metal bonds between six atoms can be realized in a prism as shown in Fig. 9.1.

9.2.2 MO Theory

The decacarbonyldimanganese complex serves as an example for a simple introduction to the chemical bond between metal centers. We start again with the monomeric fragments without metal-metal bond. To represent the pentacoordinated complex fragment, we assume an octahedral complex with one ligand removed along the z axis. We thus arrive at a square pyramidal coordination environment for the manganese with a splitting of the d orbitals as shown in Fig. 9.2. For the sake of clarity, we restrict ourselves to considering the "ligand field orbitals"; in the introduction to MO theory, we saw that this approach is quite justified and sufficient in this case.

Each manganese has seven d electrons which, starting from a low-spin state, are distributed among the five orbitals. The unpaired electron is located in the d_{z^2} orbital, which is ideally suited for the formation of a σ bond between the two metal centers. The energetically more favorable σ orbital is occupied by the two now paired electrons, while the σ^* orbital remains empty. This very illustrative observation explains why the metal-metal bond is stable (energy gain) and why the dinuclear complex is diamagnetic. However, the dissociation energy for breaking the metal-metal bond is much smaller than the C-C bond energy of ethane. The bond is significantly weaker.

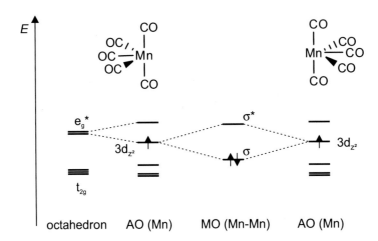

Fig. 9.2 Binding ratios in the [Mn$_2$(CO)$_{10}$]. In both monomeric fragments, the d$_{z^2}$ orbital is single occupied in each case. The lower-lying t_{2g} orbitals are all fully occupied and not relevant for the formation of a metal-metal bond. The same applies to the empty d$_{x^2-y^2}$-orbital. The d$_{z^2}$ orbital has just the right symmetry for the formation of a σ bond along the z axis. A bonding (σ) and antibonding (σ^*) molecular orbital is formed, and the energetically lower-lying bonding molecular orbital is occupied by the two electrons from the atomic orbitals

In addition to the question of whether there is a metal-metal bond, there is also the question of the formal metal-metal bond order. This is determined by the d electron number of the fragments and can be determined – as with other diatomic molecules – by counting the electrons in the bonding and antibonding molecular orbitals. Similar to carbon chemistry (alkanes, alkenes, alkynes), single, double, and triple bonds are possible here. We have already seen that metal centers are capable of forming π bonds in their complexes with π donor or π acceptor ligands. In the following we will learn that even higher bond orders are possible with metal-metal bonds.

9.3   Multiple Metal-Metal Bonds

In order to find an approach to this topic, the two dinuclear complexes [Cr$_2$(ac)$_4$(H$_2$O)$_2$] and [Cu$_2$(ac)$_4$(H$_2$O)$_2$] are compared as a first example. The molecular structure of the two acetates is shown schematically in Fig. 9.3. Both complexes exist as a dimer, with a Jahn-Teller distorted metal center (copper(II) ion = d^9 and chromium(II) ion = d^4). Copper acetate has a blue color typical for copper(II) ions and is paramagnetic at room temperature, becoming diamagnetic at lower temperatures. Chromium acetate, unlike the blue to green paramagnetic mononuclear chromium(II) complexes, is red and diamagnetic, like all dimeric chromium(II) compounds. Initial evidence of metal-metal bonding is provided by the metal-metal spacing obtained from X-ray structural analysis. This is longer in the case of copper acetate (2.64 Å) than the Cu-Cu distance in metallic copper (2.56 Å), while

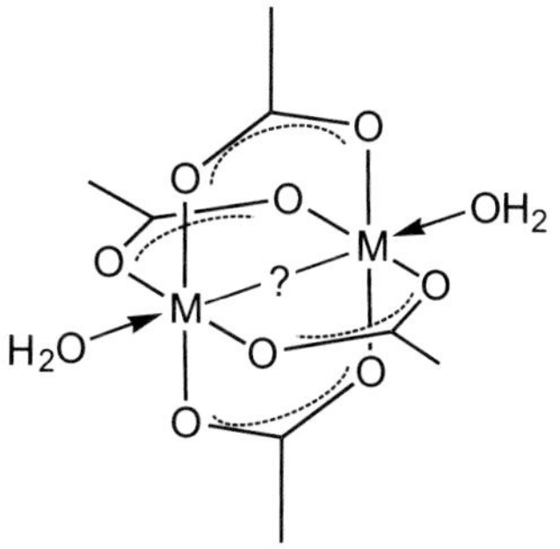

Fig. 9.3 Schematic representation of the structure of copper acetate monohydrate and chromium acetate monohydrate

in the case of chromium acetate the Cr-Cr distance of 2.36 Å is significantly shorter than the corresponding distance in metallic chromium (2.58 Å).

For the example $[Mn_2(CO)_{10}]$ it has already been pointed out that the manganese-manganese bond is responsible for the fact that the complex is diamagnetic. By forming a σ bond between the two d_{z^2} orbitals, which are each occupied by an unpaired electron, the two electrons can be arranged paired in the then energetically lower bonding orbital. The chromium acetate is also diamagnetic – in contrast to the paramagnetic mononuclear chromium complexes. And the copper acetate is also diamagnetic at low temperatures – is there a metal-metal bond between the d-orbitals in both examples?

To answer this question, we must first consider which d orbitals are occupied by unpaired electrons. In both cases, the metal center is present in a Jahn-Teller distorted octahedral ligand field and the acetate ion is a weak-field ligand. The chromium(II) ion has four valence electrons, while the copper(II) ion has nine. Consequently, the d_{xy}, d_{xz}, d_{yz} and the d_{z^2} orbital of the chromium acetate are singly occupied, and the $d_{x^2-y^2}$ orbital of the copper acetate is single occupied. Here, at most one single bond could be formed, while for chromium acetate there are four possible orbitals that can overlap. For a σ bond, the $d_{x^2-y^2}$ orbital in copper acetate does not have the correct symmetry, since the orbital lobes point to the four bridging acetate ligands. In agreement with the paramagnetic properties at higher temperatures, the long metal-metal distance and the color comparable to other monomeric complexes, no metal-metal bond can be formulated here. The magnetic properties (diamagnetic at low temperatures) must be explained by another mechanism (see magnetism/super-exchange).

In the case of chromium acetate, the situation is different. In this compound, which has been known since 1844, the Cr-Cr distance is significantly shorter than the corresponding distance in metallic chromium. A whole series of similar dimeric chromium(II) complexes are known in which a Cr-Cr distance of the order of 2.3–2.5 Å is found. In the gas phase, a Cr-Cr distance of 1.97 Å was determined for anhydrous chromium acetate. Similar spacings (1.98 Å) are determined for analogue compounds with a reduced ligand supply ($[Cr(Me)_4]_2$). This together with the very different properties of monomeric and dimeric chromium(II) complexes can be explained (in agreement with the short Cr-Cr distance) by a quadruple bond between the two Cr atoms. Between the four singly occupied d-orbitals, four bonding and four antibonding molecular orbitals are formed.

The first unequivocal demonstration of a quadruple bond between two transition metals was achieved in 1964 by Cotton et al. by crystal structure analysis on the compound $K_2[Re_2Cl_8]\cdot 2\ H_2O$. A remarkably small Re-Re distance of 2.24 Å was determined for this compound (compared to 2.75 Å in Re metal) and it has since been considered a prototype of complexes with multiple bonds between two transition metals.

Bonding In contrast to copper acetate, the d orbitals in chromium acetate are highly capable of forming metal-metal bonds. Results of crystal structure analysis show that the z axis is the bonding axis and also the Jahn-Teller axis (stretched octahedron). Thus, the d_{z^2} orbital has the suitable symmetry to form a metal-metal σ bond. Furthermore, two π bonds can be formed, respectively between the d_{xz} and the d_{yz} orbitals. This leaves only the single occupied d_{xy} orbital. Due to the very short chromium-chromium distance, a bond is also formed here in which each of the four orbital lobes overlaps with that of the opposite orbital. In this bond, there are two nodal planes along the internuclear axis and it is called the δ bond. In Fig. 9.4, the MO scheme of chromium acetate is shown. The structure of the bonding molecular orbitals formed in this process is shown on the right side of Fig. 9.4.

The metal-metal quadruple bond is also in agreement with the EAN rule. Counting the electrons per chromium center leads to a completed 18-electron shell per chromium atom with 4 d electrons, 4 Cr-Cr bond electron pairs, and 5 electron pairs from the ligands. Another way to describe the bonding is provided by the theory of localized molecular orbitals. In this model, 6 hybrid orbitals are formed per chromium center from the $3d_{x^2-y^2}$, $3d_{z^2}$, $4s$, $4p_x$, $4p_y$ and $4p_z$ orbitals. Five of these hybrid orbitals overlap with the five ligand orbitals, and the sixth hybrid orbital forms the σ bond between the two chromium centers. The two π bonds are formed between the unhybridized d_{xz} and d_{yz} atom orbitals and the interaction between the d_{xy} orbitals leads to the δ-bond. The six hybrid orbitals are arranged octahedrally around the chromium center. This means that the ligands adopt an ecliptic conformation rather than the sterically favorable staggered one. This is not surprising in the case of chromium acetate with the bridging acetate ligands – but the same conformation is also found in the case of $[Re_2Cl_8]^{2-}$. A look at the δ bond shown in Fig. 9.4 makes it clear that this bond can only be formed when the ligands are in an ecliptic arrangement. Excititing the δ-δ^* transition leads to an excited state with a staggered conformation. This transition is responsible for the red color of the dimeric chromium compounds.

The description of the bonding as $(\sigma^2)(\pi^4)(\delta^2)$-quadruple bonds is a simple and very descriptive representation. However, it only reflects reality to a limited extent, since it assumes that each of the four bonding molecular orbitals is doubly occupied by two electrons. A bond analysis of $[Re_2Cl_8]^{2-}$ with CASSCF methods (complete active space self-consistent field) yields a calculated Re-Re bond order of 3.2. The contribution of the δ

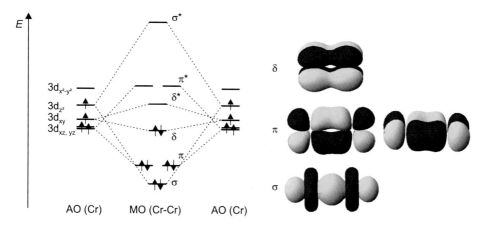

Fig. 9.4 Left: MO scheme for the M-M quadruple bond in chromium acetate. Right: Structure of the four binding molecular orbitals [18]

bond is about 0.5. The decisive factor is the partially occupied antibonding δ^* state. Now it is a question of the definition of "bond" and "bond order". The bond could be described as a weak quadruple bond with four orbitals of bond-relevant overlap, or as a bond with four pairs of electrons and the effective bond order of about 3. The bond order is determined by counting the electrons in bonding and antibonding molecular orbitals.

9.3.1   Higher, Stronger, Shorter: Metal-Metal Quintuple Bond

Based on the previous considerations, a metal-metal quintuple bond is also conceivable. For this, one would have to reduce the oxidation state of the chromium in a binuclear chromium compound from +2 to +1, for example. The additional electron occupies the previously empty $d_{x^2-y^2}$ orbital and in this way a fivefold overlap of the metal d orbitals is possible. Such a bond could also be called a ten-electron two-center bond. Figure 9.5a shows the structure of the first compound for which such bonding has been discussed [38]. In this compound, as in the following ones, a sterically very demanding ligand was used to shield the Cr-Cr bond in [ArCrCrAr], with Ar = monoanionic aromatic ligand, thus protecting it from subsequent reactions. Analogous to the previous example, although five pairs of electrons are involved in the bond, the bond order is much lower than 5, as the higher antibonding states are also partially occupied in this case. This is reflected in the strikingly long Cr-Cr distance of 1.835 Å, which is longer than the shortest distances discussed for Cr-Cr quadruple bonds. Thus, even shorter bonds are conceivable here. Another striking feature is the angled arrangement of the compound, which is partly due to stabilizing metal-aromatic interactions (thin line). Another reason for the *trans*-angled structure of the compound, according to DFT studies, are strong σ bonds of the $4s3d_{z^2}$ hybrid orbitals [39].

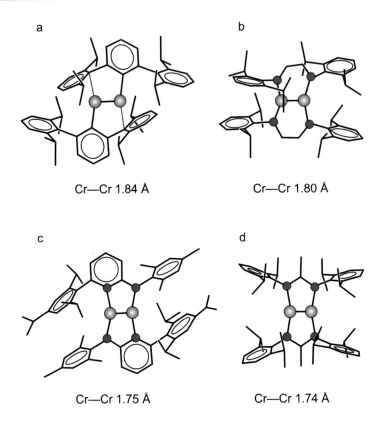

Fig. 9.5 (**a**) Structure of the first stable dimeric chromium compound in which a quintuple bond was discussed [38]. (**b–d**) Structure of dimeric chromium compounds with ultrashort chromium-chromium quintuple bonds (in order of appearance) [40]

Even shorter Cr-Cr bonds were achieved by clever choice of bridging ligands. The shortest Cr-Cr distances of 1.75 Å and 1.74 Å observed so far (both published simultaneously in 2008) were achieved with triatomic bridging ligands. The Cr-Cr distances are intermediate between those discussed for chromium(II) complexes with quadruple bonds and the distance of 1.68 Å observed in the gas-phase molecule Cr_2, which is due to a thinkable sextuple bond (see below). The very short Cr-Cr bonds are due to the sterically demanding ligands. The calculated bond order of compound c in Fig. 9.5 is 4.2, which corresponds to a formal quintuple bond. The fact that the value is significantly less than 5 is due to the weak δ bonds [40].

If the idea of minimizing the number of ligands to increase the oxidation state and thus also the bond order, which was started with chromium acetate, is pursued further, a sextuple bond could be realized for dimeric $[Cr_2^0]$, where a further σ bond is formed by the interaction between the 4s orbitals. By minimizing the number of ligands and reducing the oxidation state, the bond order is further increased continuously. Indeed, the Cr_2

molecule can be prepared and characterized in the gas phase and the shortest distance determined in the process is 1.68 Å, which is once again significantly shorter than that of the compounds with fivefold bonds. However, the $(4s\sigma^2)(3d\sigma^2)(3d\pi^4)(3d\delta^4)$-sixfold bond remains rather hypothetical in nature. Calculations show that strong 3d interactions are present at short metal-metal distance, while at larger distance the 4s interaction dominates.

9.4   Cluster Complexes

Having discussed the structure and bonding of dinuclear complexes in detail, we return to the multinuclear systems, the cluster complexes. The EAN rule can be used to predict the number of metal-metal bonds in many carbonyl complexes, and thus conclusions can be drawn about the structure of the complexes. However, there are also examples where this simple relationship fails. An example for this is the complexes $[Os_6 (CO)_{18}]$ and $[Os_6(CO)_{18}]^{2-}$ given in Fig. 9.6. For the first complex, we would expect twelve metal-metal bonds using the EAN rule. The simplest polyhedron that can be used to realize this number of bonds between six atoms is an octahedron. The actual structure of the complex does show that there are 12 Os-Os bonds, but the structure is not an octahedron, but a bicapped tetrahedron. The corresponding dianion has two more electrons in the electron count and thus it should have only 11 metal-metal bonds according to the EAN rule. However, 12 M-M bonds are also observed, this time the structure is the octahedron. Calculations could be performed to predict the structure of such cluster complexes, but they are time-consuming on such large molecules. Another simple bonding concept based on electron counting rules to predict the framework structure of carbonyl complexes would be ideal.

The EAN rule assumes localized electron pairs; in the present example, especially for the negatively charged complex, the electrons are delocalized. A different model is needed to predict the framework structure. The isolobal analogy shows us that the BH fragment of boranes is isolobal to an $M(CO)_3$ fragment of cluster complexes (with M = Fe, Ru, Os). For this reason, the structure of carbonyl cluster complexes of these metals can be predicted using the same rules used to predict the structure of polyboranes, the Wade rules. In the following, we first explain the Isolobal principle and the Wade rules before applying our findings to the example of carbonyl cluster complexes mentioned in the introduction.

9.4.1   The Isolobal Analogy

The isolobal analogy was introduced by Roald Hoffmann, who was awarded the Nobel Prize in Chemistry for it in 1981 [41, 42]. What is special about this concept is that it bridges the gap between organic and inorganic chemistry. We start with a definition:

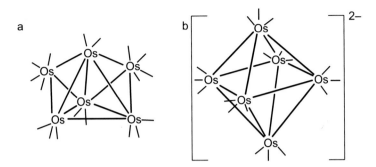

Fig. 9.6 Structure of $[Os_6(CO)_{18}]$ (**a**) and $[Os_6(CO)_{18}]^{2-}$ (**b**). For reasons of clarity, only the Os atoms, the Os-Os bonds and the bonds to the carbonyl ligands are shown

The Isolobal Analogy

Molecular fragments are isolobal if the number, symmetry properties, approximate energy and shape of the frontier orbitals, and the number of electrons in these orbitals are similar.

A good introduction to this concept is obtained by looking at the number of electrons required for molecular fragments or atoms to reach the noble gas configuration. This relationship is illustrated in Fig. 9.7. The chlorine atom, CH_3 radical, and $[Mn(CO)_5]$ fragment are mutually isolobal. All of them are still missing one electron for reaching the noble gas configuration and the corresponding frontier orbitals are similar. The symbol for isolobality is a two-headed arrow with half an orbital below it. By forming a σ bond between two molecular fragments or atoms, the electron deficiency is remedied. This can also involve linking two different fragments that are isolobal to each other. All examples shown in Fig. 9.7 are known, stable compounds.

This principle can now be applied to atoms or molecular fragments which are lacking two or three electrons to reach the noble gas configuration. Corresponding examples are given in Fig. 9.8.

That the number and energy of the frontier orbitals are similar for the examples shown here can be illustrated with the aid of the molecular orbital diagrams of the corresponding molecular fragments. As an example, the MO scheme of a CH_3 fragment and an ML_5 fragment (where M is a metal with seven valence electrons) are given in Fig. 9.9. Both have a semi-occupied frontier orbital (orbital with unpaired electron) and the frontier orbitals have similar energy.

The molecular fragments considered so far have always been neutral compounds. In the examples in Fig. 9.8 we have already seen that osmium (8 valence electrons) can simply be exchanged for iron (also 8 valence electrons). The whole concept can now easily be applied to charged species. We start again with the $[M(CO)_5]$ fragment with 17 valence electrons, which is isolobal to CH_3. Other complex fragments with 17 valence electrons and five

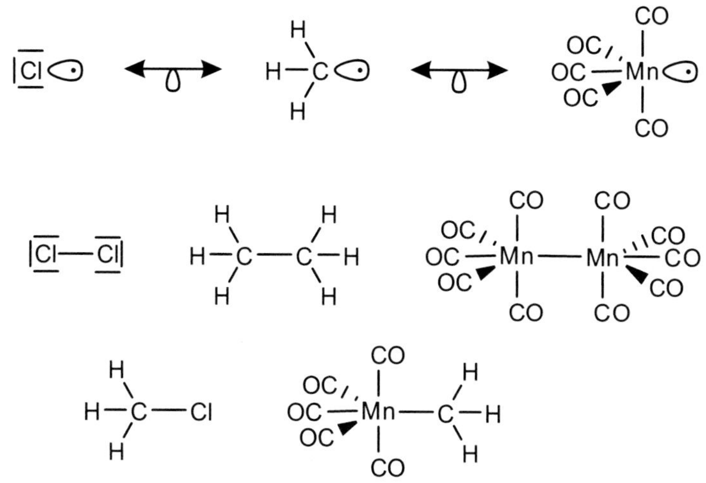

Fig. 9.7 Isolobal relationship between the chlorine atom, the methyl radical and the $[Mn(CO)_5]$ fragment. The common feature in these three examples is that one electron is still missing until the noble gas configuration is reached. The chlorine atom and the methyl radical have 7 valence electrons, while the $[Mn(CO)_5]$ fragment has 17. The symbol used for isolobality is a two-headed arrow with half an orbital below it. By forming a single bond between two fragments, both fragments reach the desired noble gas configuration. The compounds formed in this process are all known

ligands would be $[Fe(CO)_5]^+$ or $[Cr(CO)_5]^-$. Both are still missing one valence electron to the closed noble gas shell and are therefore isolobal to the CH_3 radical. In the same way, the charge of the organic fragment can be varied. CH_3^+ is a 6 valence electron fragment which is isolobal to complex fragments with five ligands and 16 valence electrons such as [Cr $(CO)_5$]. Correspondingly, CH_3^- is then an 8 valence electron fragment and isolobal to 18 valence electron complexes with five ligands such as $[Fe(CO)_5]$.

This leads to the connection between cluster complexes of carbonyls and polyboranes mentioned at the beginning. The BH fragment has the same number of valence electrons as a CH^+ fragment, both are isolobal with 4 valence electrons each. Corresponding complex fragments with 14 valence electrons and three ligands are the $M(CO)_3$ fragments with M = Fe, Ru, Os already mentioned at the beginning. Accordingly, the structure of cluster complexes from these fragments should be comparable to the structure of polyboranes. To understand and predict these, we need the Wade rules.

9.4.2 The Wade Rules for Boran Clusters

The Wade rules for predicting structures of polyboranes (electron deficient compounds) were established by K. Wade, R. E. Williams, and R. W. Rudolph. For the generation of the structures we assume that the boron atoms occupy the vertices of polyhedra bound only by

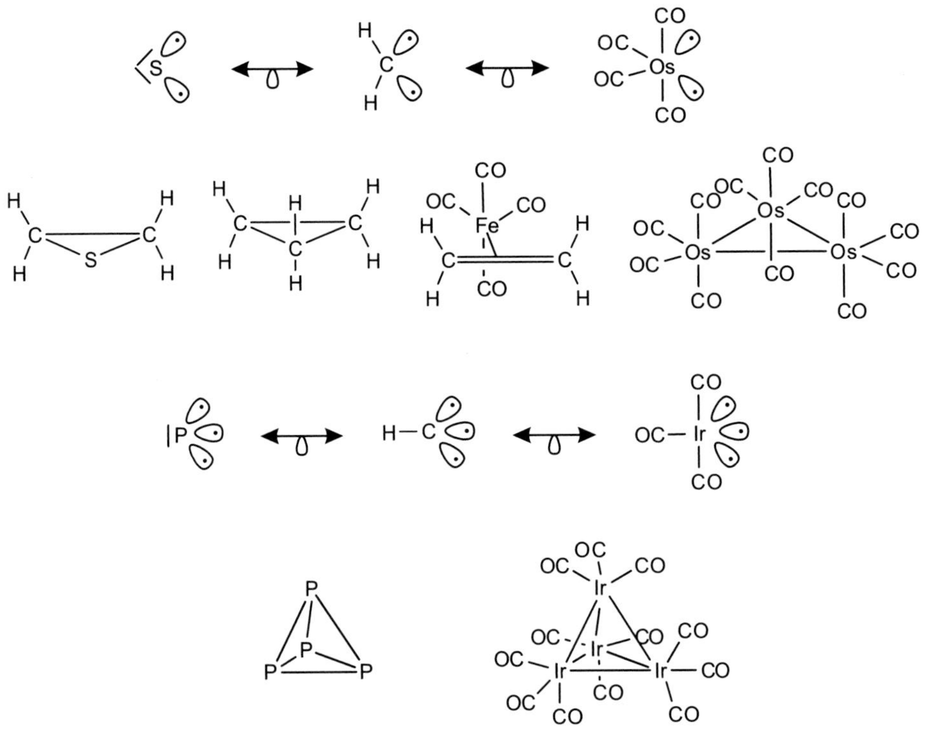

Fig. 9.8 Examples of molecular fragments or atoms that are insulobal to each other and lack two or three electrons to reach the noble gas configuration. The compounds derived from these fragments are all known to exist

triangular faces. Accordingly, the tetrahedron (4 boron atoms), the trigonal bipyramid (5 boron atoms), the octahedron (6 boron atoms), the pentagonal bipyramid (7 boron atoms), etc. occur as polyhedra. The boranes of the general composition B_nH_m can be divided into different types according to the ratio of boron to hydride, the first four of which are mentioned here.

- *closo-boranes* have the general composition B_nH_{n+2}. In these boranes, all corners of the polyhedron are occupied by boron atoms.
- *nido-boranes* have the general composition B_nH_{n+4}. In these boranes, all but one of the corners of the polyhedron are occupied by boron atoms.
- *arachno-boranes* have the general composition B_nH_{n+6}. In these boranes, all but two corners of the polyhedron are occupied by boron atoms.
- *hypho-boranes* have the general composition B_nH_{n+8}. In these boranes, all but three corners of the polyhedron are occupied by boron atoms.

This classification can be continued on and on. On the outside of the boron skeleton, each boron atom has a terminal H atom, also known as an "exo" hydrogen atom. The other

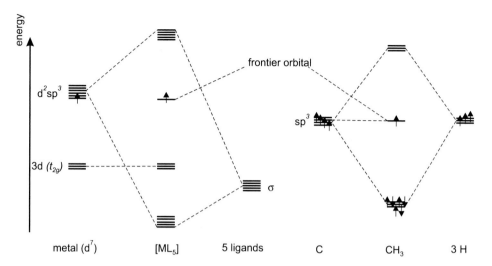

Fig. 9.9 MO scheme of an ML_5 fragment (where M is a metal with seven valence electrons) and a CH_3 fragment. The orbitals energetically below the frontier orbital are fully occupied with electrons. For the sake of clarity, we did not draw the electron pairs in the MO scheme for the ML_5 fragment. In this consideration, only σ bonds are taken into account and it is assumed that hybrid orbitals are formed for the carbon (sp^3) and the metal (d^2sp^3)

hydrogen atoms form three-center two-electron bonds. For these bonds, three atom orbitals, only two of which are singly occupied while the third is empty, overlap to form three molecular orbitals in which the lowest-energy bonding molecular orbital is occupied by the two electrons. In Fig. 9.10, a *closo,* a *nido,* and an *arachno* borane are shown as examples. Three-center two-electron bonding is illustrated on the right using B_2H_6 as an example. In the boranes, the bond hydrogen atom corresponds to a hydride. This means that a negative charge of the polyborate is counted like a hydrogen.

9.4.3 The Wade-Mingos Rules

With the help of Mingos' rules (or polyhedron-skeletal electron pair (PSEP) theory or Wade-Mingos rules), the very illustrative Wade rules can also be transferred to other cluster complexes and by this be used to predict the structure of carbonyl complexes. To do this, we have to count electrons. As for the 18-electrons rule, only the valence electrons, or in this case the skeletal electrons, are important. The number of skeletal electrons in a cluster relative to the number of atoms determines which structure is taken.

- If a cluster with n atoms has $2n + 2$ skeletal electrons, then it has a *closo structure.*
- If a cluster with n atoms has $2n + 4$ skeletal electrons, then it has a *nido structure.*
- If a cluster with n atoms has $2n + 6$ skeletal electrons, then it has an *arachno structure.*
- If a cluster with n atoms has $2n + 8$ skeletal electrons, then it has a *hypho-structure.*

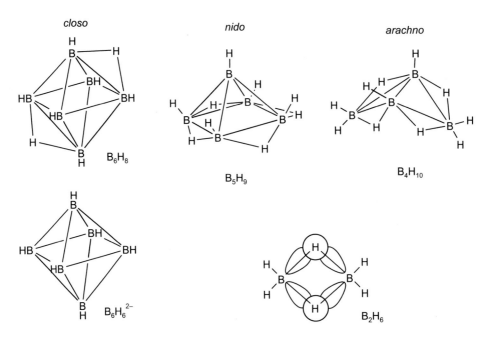

Fig. 9.10 Structure of polyboranes

Now only the electron counting rules are needed. The valence electrons from the metal (v) and the valence electrons from the ligands $(l$, per ligand, for $H = 1$, all Lewis bases $= 2$) are counted. For main group compounds such as the boranes, from the sum of $v + l$ (Wade rules) per fragment (or per boron atom) two is subtracted, and 12 is subtracted for the d-block compounds (extension of Wade rules, Wade-Mingos rules). Some examples follow to illustrate the concept.

The first examples come from the boranes. These are made up of BH fragments and BH_2 fragments. Boron has three valence electrons, and hydrogen one. Thus for a BH fragment $v + l - 2 = 3 + 1 - 2 = 2$. For a BH_2 fragment we get $v + l - 2 = 3 + 2 - 2 = 3$. The compound B_6H_8 (or $B_6H_6^{2-}$) is built from 4 BH fragments and 2 BH_2 fragments. This gives $4 \cdot 2 + 2 \cdot 3 = 14$ skeletal electrons. Since we have 6 boron atoms, this corresponds to $2n + 2$ skeletal electrons and we expect a *closo* framework structure as also seen in Fig. 9.10. The compound $B_5 H_9$ is composed of one BH fragment and 4 BH_2 fragments. This gives $1 \cdot 2 + 4 \cdot 3 = 14$ skeletal electrons. Since we have 5 boron atoms, these are $2n + 4$ skeletal electrons and we expect a *nido* framework structure. The last example is B_4H_{10}, which is built from 2 BH_2-fragments and 2 BH_3-fragments. Each BH_3 fragment contributes 4 skeletal electrons. Thus, we obtain a total of 14 skeletal electrons, which corresponds to an *arachno* structure with 4 boron atoms.

The last two examples are the carbonyl cluster complexes mentioned at the beginning. As a first example, we consider the cluster complex $[Os_6(CO)_{18}]^{2-}$, for which the EAN rule had completely failed. This cluster complex is composed of 4 $M(CO)_3$ fragments and 2 M $(CO)_3^-$ fragments. For each of these units, the osmium provides 8 valence electrons, the carbonyl ligands each provide two valence electrons, and each negative charge counts as one valence electron. To determine the number of skeletal electrons, we must subtract 12 from the sum. We obtain $8 + 3 \cdot 2 - 12 = 2$ for the $M(CO)_3$ fragments and 3 for the M $(CO)_3^-$ fragments. Thus, the cluster complex $[Os_6(CO)_{18}]^{2-}$ has 14 skeletal electrons, which corresponds to the formula $2n + 2$. We obtain a *closo* structure and for $n = 6$ this corresponds to an octahedron. This result agrees excellently with the experimentally determined structure! The second cluster complex, $[Os_6(CO)_{18}]$, is composed of 6 M $(CO)_3$ fragments. This gives us 12 valence electrons, which corresponds to $2n$. The Wade-Mingos rules fail. Whenever this is the case, capped polyhedra are observed, whose structure can be predicted by the capping rule, which is not discussed further here.

9.5 Questions

1. Using the EAN rule, try to determine the number of metal-metal bonds for the following clusters. Can the rule be applied to all clusters? What polyhedra do you expect for the metal clusters? $[Ru_4(CO)_{12}H_2]^{2-}$, $[Re_4H_4(CO)_{12}]$, $[Ta_6(\mu\text{-}Cl)_{12}Cl_6]^{4-}$, $[Ir_4(CO)_{12}]$, $[Ni_8(PPh_3)_6(CO)]_8^{2-}$
2. Discuss the bonding in $[Re_2(CO)_{10}]$ and $[Re_2Cl_8]^{2-}$ using the EAN rule and MO theory!
3. What conditions must be present for a metal-metal quadruple or quintuple bond to be realized? Discuss the molecular orbitals involved in the bond!
4. What is the difference between σ, π and δ bonds?
5. Roald Hoffmann was awarded the Nobel Prize in Chemistry in 1981 for introducing the isolobal analogy. Which of the following molecular fragments are isolobal to each other? Give reasons for your decision! CH_3^-, BH, CH_3^+, $[Mn(CO)_5]$, $[Fe(CO)_5]$, $[Ru(CO)_3]$, $[Ir(CO)_3]$

Magnetism

10

A special feature of the transition metals (d elements), but also of the lanthanides and actinides (f elements), compared to the main group elements is the possibility to realize different oxidation states. Many of these compounds have unpaired electrons in the d or f orbitals. These are involved in bonding, especially in the case of the d orbitals, and are influenced by ligands. This means that the coordination environment around the metal center affects its optical and magnetic properties. In other words, the magnetic properties of a compound allow conclusions to be drawn about its electronic structure. An example of this from coordination chemistry are nickel(II) complexes. With four strong-field ligands such as cyanide, the complex geometry is square-planar and the complexes are diamagnetic. In contrast, a corresponding complex with four weak-field ligands such as chloride is paramagnetic and the coordination geometry is tetrahedral. Using ligand field theory, we can explain the different magnetic properties of the two complexes. Because of the close connection between complexes and magnetism, this chapter is first devoted to some general aspects of the phenomenon of magnetism, before specifics from the field of coordination chemistry are considered in more detail. Similar to the section on the color of coordination compounds, parallels with solid-state chemistry can be identified and will be pointed out at the appropriate place.

The magnetic properties of compounds are not only of interest for a better understanding of the electronic structure, they are becoming more and more important as a material property. Magnetic materials occur in many areas of everyday life. We encounter them as permanent magnets (e.g. the "classical" magnets, hard magnetic materials), in storage media and transformer sheets (soft magnets) or as switches in high frequency technology (non-conducting oxides). Magnetic materials can be metals (Fe, Co, Ni, Gd, . . .), alloys ($SmCo_5$, Heusler phases such as Co_2MnAl, Co_2CrAl, Co_2FeSi) or oxides (CrO_2, Fe_3O_4, $ZnFe_2O_4$). If the materials are made up of molecules, as in the case of coordination

B. Weber, *Coordination Chemistry*,
https://doi.org/10.1007/978-3-662-66441-4_10

compounds, for example, we also speak of molecular magnets and molecular magnetism. The special attraction for this class of compounds lies in the possibility of achieving properties that are not possible for classical solids, e.g. transparency or flexibility. In the following section, we will learn where magnetism comes from, what types of magnetism there are, how it can be characterized and – this is particularly important for materials – how specific magnetic properties can be realized on purpose.

10.1 Units

A book chapter or lecture on magnetism cannot begin without briefly turning to units. Most scientists working in the field of molecular magnets use the so-called CGS-emu system in place of the SI system of units recommended by IUPAC. While the SI system is based on the four basic quantities length (m), mass (kg), time (s) and current (A), the Gaussian CGS system (or CGS-emu system, emu stands for electro-magnetic unit) uses the three mechanical quantities length (cm), mass (g) and time (s). Conversions between the two competing systems are often necessary for everyday magnetochemical work. This is complicated by the fact that, in addition to powers of 10, the factor 4π is often included. Table 10.2 summarizes the most important magnetic quantities and their units with conversion factor. Table 10.1 shows some frequently used physical constants (both taken from "Practical Guide to Measurement and Interpretation of Magnetic Properties (IUPAC Technical Report)" [43] (Table 10.2).

IUPAC recommends the use of quantities whose values are independent of the system used. This includes the number of effective Bohr magnetons μ_{eff}. Although the volume susceptibility χ is unitless, it is converted by a factor of 4π. The same is true for the magnetic field strength. To avoid confusion, it is recommended to use the "magnetic field" B_0 (the external magnetic induction) in diagrams, where the conversion factor between units is 10^{-4} T/G.

Table 10.1 Physical constants; taken from "IUPAC Technical Report" [43]

Symbol		SI	CGS-emu
h	Planck's constant	6.62607×10^{-34} J s	6.62607×10^{-27} erg s
k_B	Boltzmann constant	1.38066×10^{-23} J/K	1.38066×10^{-16} erg/K
μ_B	Bohr magneton	9.27402×10^{-24} A m^2	9.27402×10^{-21} G cm^3
c_0	Speed of light in vacuo	2.99792458×10^{8} m/s	$2.99792458 \times 10^{10}$ cm/s
m_e	Mass of the electron	9.10939×10^{-31} kg	9.10939×10^{-28} g
N_A	Avogadro constant	6.02214×10^{23} mol^{-1}	
e	Elementary charge	1.60218×10^{-19} C	

Table 10.2 Definitions, units [] and conversion factors; taken from "IUPAC Technical Report" [43]

	Size	SI	CGS-emu	Factor
μ_0	Magnetic field constant	$\mu_0 = 4\,\pi\times 10^{-7}$ [Vs/Am]	1	
B	Magnetic induction	$B = \mu_0\,(H + M)$ [T = V s/m²]	$B = H^{(ir)} + 4\pi\,M$ [G]	10^{-4} T/G
H	Magnetic field strength	H [A/m]	[Oe]ᵃ	10^3 Oe/4π A/m
B_0	"Magnetic field"	$B_0 = \mu_0\,H$ [T]	[G]ᵃ	10^{-4} T/G
M	Magnetization	M [A/m]	[G]ᵃ	10^3 (A/m)/G
μ_B	Bohr magneton	$\mu_B = eh\,/2m_e$ [A m²]	$\mu_B = eh\,/2m_e$ [G cm³]	10^{-3} A m²/G cm³
χ	Mag. volume susceptibility	$M = \chi\,H$	$M = \chi^{(ir)}\,H^{(ir)}$	4π
χ_g	Mag. gram susceptibility	$\chi_g = \chi/\rho$ [m³/kg]	$\chi_g = \chi^{(ir)}/\rho^{(ir)}$ [cm³/g]	$4\pi\,/10^3$ (m³/kg)/(cm³/g)
χ_M	Mag. molar susceptibility	$\chi_M = \chi M/\rho$ [m³/mol]	$\chi_M = \chi^{(ir)}M/\rho^{(ir)}$ [cm³/mol]	$4\pi\,/10^6$ m³/cm³
μ_{eff}	Bohr's magneton number	$[3k_B/\mu_0 N_A\,\mu_B^2]^{1/2}\ [\chi_m T]^{1/2}$	$[3k_B/N_A\mu_B^2]^{1/2}\ [\chi_m T]^{1/2}$	

ᵃThe use of Gauss and Oersted seems somewhat arbitrary in the CGS-emu system. Since the magnetic field constant here is $\mu_0 = 1$, $1\,G = 1\,Oe$

10.2   Magnetic Properties of Matter

In order to derive the relationship between the individual quantities used in the field of magnetism, we first consider how magnetic fields are generated in physics. For this we need the coil given in Fig. 10.1, through which an electric current (direct current) flows. Each coil through which current flows generates a magnetic field. The magnetic field strength H depends on the length of the coil L, the number of windings n and the current strength I according to:

$$|H| = \frac{n \cdot I}{L}$$

The magnetic induction or flux density B is a measure of the density of the field lines, i.e. the strength of the magnetic field. In matter-free space (vacuum) it is calculated as:

$$B = \mu_0 \cdot H$$

Here μ_0 is the magnetic field constant (permeability) in a vacuum. The magnetic induction inside a body is different from that in a vacuum. The change ΔB is called magnetic

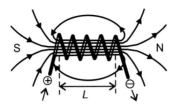

Fig. 10.1 A coil with current flowing through it generates a magnetic field. The field strength depends on the length of the coil, the number of windings and the current strength. N and S stand for the north pole and the south pole of the magnet

polarization B_{pola} or magnetization M. Depending on the material we consider, the magnetization can be positive or negative. Instead of magnetization, the volume susceptibility χ_V is usually used as a measure of the magnetization of a sample. Susceptibility means "ability to absorb lines of force". The volume susceptibility is a unitless proportionality factor.

$$B_{\text{inside}} = B_{\text{outside}} + B_{\text{pola}} = \mu_0(H + M) = \mu_0 H(1 + \chi_V)$$

with

$$\chi_V = \frac{M}{H}$$

The magnetic permeability of a substance μ_r is also given as a substance quantity. It is related to the previously mentioned quantities as follows.

$$B_{\text{inside}} = \mu_r \cdot B_{\text{outside}} = \mu_r \cdot \mu_0 \cdot H$$

with

$$\chi_V = \mu_r - 1$$

All methods for susceptibility measurement are based on the determination of the quotient $\frac{B}{H}$. Instead of the volume susceptibility, the magnetic susceptibility per gram ($\chi_g = \frac{\chi_V}{\rho}$) or per mole ($\chi_M = \chi_g \times M$) is usually used. For the following considerations, the molar magnetic susceptibility plays a role, which is denoted by χ for the sake of simplicity.

There are two different types of magnetic behavior, which are distinguished according to the sign of M or χ_V; diamagnetism and paramagnetism. In Fig. 10.2 the course of the field lines for both variants is given. In addition, there are a number of cooperative magnetic phenomena such as ferromagnetism, to which we will return later. Table 10.3 summarizes the differences of the volume susceptibility and the magnetic permeability depending on the magnetic behavior of the matter.

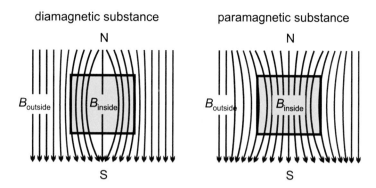

Fig. 10.2 Course of the field lines of an external magnetic field in the presence of a diamagnetic or paramagnetic substance

Table 10.3 Order of magnitude of volume susceptibility and magnetic permeability as a function of magnetic properties

Magnetism	χ_V	μ_r	Change with T
Dia-	$<0(\approx 10^{-5})$	<1	–
Para-	$>0(0-10^{-2})$	>1	Decreases
Ferro-	$>0(10^{-2}-10^{6})$	>1	Decreases

10.2.1 Diamagnetism

Diamagnetism is observed for compounds that have exclusively paired electrons. For such compounds, the magnetization is negative. The magnetic field line density in the body is lower than outside, or in an inhomogeneous magnetic field the body moves to the area with weaker magnetic field ("repulsion"). The magnitude of molar susceptibility is in the range of -1 to -100×10^{-6} emu/mol and is independent of field strength and temperature. Diamagnetism is a general property of matter caused by the interactions of electrons (moving charge) with the external magnetic field. In order to analyze the "pure" paramagnetism of a compound, it is therefore important to correct the measured susceptibility of a material by its diamagnetism. Especially in complexes with large organic ligands, the diamagnetism of the ligand provides a crucial contribution to the observed magnetic moment [44].

$$\chi = \chi^P + \chi^D$$

There are several ways to determine diamagnetic susceptibility, three of which are listed here [44]:

- Application of the equation $\chi^D = -k \times M \times 10^{-6}$ *emu/mol* to estimate the diamagnetism of ligands and/or counterions. (The arbitrarily chosen constant k can take values between 0.4 and 0.5).
- Estimation of diamagnetism using Pascal constants (additive, Table 10.4)
- Determination of the diamagnetic susceptibility of the free ligand or the corresponding Na/K salt

The estimates are helpful for "smaller" coordination compounds, but fail if the ligands become too large in relation to the metal center (more precisely: the number of unpaired electrons), as is the case, for example, with metalloproteins. In this case, it is necessary to determine the diamagnetism as accurately as possible. The Salen ligand ($C_{16} H_{14} N_2 O_2^{2-}$, see Fig. 2.11) provides a good calculation example:

Diamagnetism measured:	-182×10^{-6} emu/mol
Equation:	-133×10^{-6} emu/mol with $k = 0.5$
Pascal constants:	-147.6×10^{-6} emu/mol

When using the Pascal constants, the corresponding values per atom are added up and corrected for the peculiarities of certain bond types (double bond, aromatic ring system).

10.2.2 Paramagnetism

Paramagnetism is observed for compounds with unpaired electrons. The magnetization is positive and the body moves towards the stronger field ("attraction") in an inhomogeneous magnetic field. The paramagnetic susceptibility is positive and the magnitude is $100–100,000 \times 10^{-6}$ emu/mol. Since the effect of paramagnetism is several orders of magnitude larger than diamagnetism, the estimation of the latter is given by the possibilities presented above. The cause of paramagnetism lies in the interaction of the spin and/or orbital moment of the unpaired electrons with the applied magnetic field. The unpaired electrons present in a paramagnet can be equated with a moving charge which, by analogy with a coil, generates a magnetic field. The magnetic moment of the unpaired electrons aligns itself parallel to the external magnetic field, resulting in an increase of the field line density.

Table 10.4 Diamagnetic susceptibilities and constitutive corrections (in 10^{-6} emu/mol; Pascal constants) taken from [44]

Atoms				Constitutive corrections			
H	−2.9	As(III)	−20.9	C=C	5.5	C=N	0.8
C	−6.0	As(V)	−43.0	C≡C	0.8	N=N	1.8
N(ring)	−4.6	F	−6.3	−C aromatic	−0.25	N=O	1.7
N(Open chain)	−5.6	Cl	−20.1	C=N	8.1	C-Cl	3.1
N(Imine)	−2.1	Br	−30.6				
O(ether, alcohol)	−4.6	I	−44.6				
O(carbonyl)	−1.7	S	−15.0				
P	−26.3	Se	−23.0				

Cations				Anions			
Li^+	−1.0	Ca^{2+}	−10.4	O^{2-}	−12.0	CN^-	−13.0
Na^+	−6.8	Sr^{2+}	−19.0	S^{2-}	−30	NO_2^-	−10.0
K^+	−14.9	Ba^{2+}	−26.5	F^-	−9.1	NO_3^-	−18.9
Rb^+	−22.5	Zn^{2+}	−15.0	Cl^-	−23.4	NCS^-	−31.0
Cs^+	−35.0	Cd^{2+}	−24	Br^-	−34.6	CO_3^{2-}	−28
NH_4^+	−13.3	Hg^{2+}	−40	I^-	−50.6	ClO_4^-	−32.0
Mg^{2+}	−5.0			OH^-	−12.0	SO_4^{2-}	−40.1

Ligands		Transition metals			
H_2O	−13	Fe^{2+}	−13	Fe^{3+}	−10
NH_3	−18	Ni^{2+}	−10		
CO	−10	Cu^{2+}	−11	Cu^+	−12
CH_3COO^-	−30	Co^{2+}	−12	Co^{3+}	−10
$C_2O_4^{2-}$	−25				
$C_5H_5^-$	−65				
Acetylacetonato	−52				
Pyridine	−49				
Pyrazine	−50				
Bipyridine	−105				
Salen	−182				
Phenantroline	−128				

10.3 The Magnetic Moment

10.3.1 Origin of the Magnetic Moment

Paramagnetism and cooperative magnetic phenomena are related to unpaired electrons. We consider the electron as a particle, and then we assume that it moves in a circular orbit around the atomic nucleus. Since the electron has a charge, the orbital motion of the electron can be equated to a conductor loop with current flowing through it. The magnetic moment induced by the current flow is the product of the area A encompassed by the coil or, since we have only one winding, by the conductor loop, multiplied by the current I.

$$m = I \cdot A$$

For a circular path, the area is $A = \pi \cdot r^2$. The current indicates how much charge flows per time. The charge of the electron is -e and the time for one orbit is $\frac{2\pi r}{v}$, where v is the velocity and r is the radius. To find the magnetic dipole moment of an electron, we need to substitute everything into the equation and add the mass of the electron m_e. The result shows that the magnetic dipole moment of an electron depends on its angular momentum $l = m_e \cdot v \cdot r$ and the factor $\frac{-e}{2m_e}$, the so-called gyromagnetic ratio γ_e of the electron.

$$\mu_e = -e \frac{v}{2\pi r} \pi r^2 = -\frac{e}{2m_e} m_e r v = \gamma_e \cdot l$$

The magnetic moment of an electron depends on its angular momentum. This in turn depends on the orbital in which the electron is located. The latter is described by the quantum numbers. For the magnetic moment (or the angular momentum of the electron), the magnetic orbital angular momentum quantum number and the spin quantum number are important. In many cases, the electron configuration is not sufficient to make a statement about which orbitals the (unpaired) electrons are in. For this we need term symbols, the determination of which was described in the chapter Binding Models.

10.3.2 Spin-Orbit and *j-j* Coupling

The magnetic moments resulting from the magnetic quantum number m_l and the spin quantum number m_s interact with each other (Fig. 10.3). There are two different interaction mechanisms for this, spin-orbit coupling (*L-S* or Russel-Saunders coupling) and *j-j* coupling. Spin-orbit coupling occurs for light atoms including 3d ions. For these elements, the coupling between spin and orbital angular momentum of an electron is weak. As a consequence, initially the spins of all electrons of an atom couple with each other and a total spin is obtained, which is characterized by the total spin quantum number S, which we already learned about when we determined the term symbols. In the same way, all orbital angular momenta couple with each other to give the total orbital angular momentum quantum number L. The coupling of L and S gives the total angular momentum for the system. In Fig. 10.3 Bottom, the *L-S* or Russel-Saunders coupling is shown schematically. The magnetic moment of a free Russel-Saunders coupled ion is determined by the quantum numbers L, S, and J of the ground state. The electron configuration of an atom or ion in the ground state is determined by Hund's rules and uniquely characterized by the corresponding term symbol. To determine the magnetic moment μ (sometimes referred

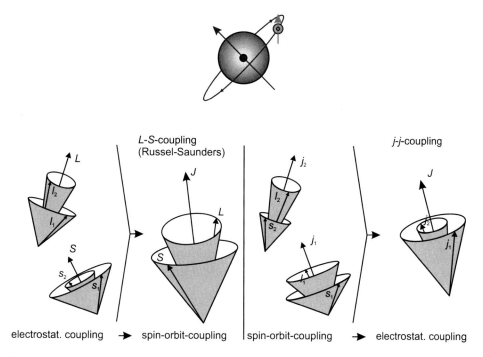

Fig. 10.3 Schematic representation of spin and orbital moment *(top)* and *L-S* (Russel-Saunders) and *j-j* coupling *(bottom)*

to as μ_a for atomic magnetic moment) in the ground state as a function of S, L and J, the equation

$$\frac{\mu}{\mu_B} = g_J \cdot \sqrt{J(J+1)}$$

with

$$g_J = 1 + \frac{S(S+1) + J(J+1) - L(L+1)}{2\,J(J+1)}$$

is used. Here, g_J is the Landé factor and μ_B is the Bohr magneton, a constant composed of the following natural constants

$$\mu_B = \frac{eh}{4\pi m_e} = 9.274 \cdot 10^{-24}\,\text{Am}^2$$

with the elementary charge e, Planck's constant h and the mass of the electron m_e. Alternatively, the Curie constant can also be calculated. It is calculated according to

$$C = \frac{\mu_0 N_A}{3k_B} \cdot \mu^2 = \frac{\mu_0 N_A}{3k_B} g_J^2 \, J(J+1)\mu_B^2$$

with the Boltzman constant k_B, the Avogadro number N_A and the permeability of the vacuum (magnetic field constant) μ_0.

For heavy elements such as ions of the lanthanides, the spin-orbit coupling is strong. In this case, the j-j coupling scheme occurs. In this coupling scheme, the angular momentum j is first determined for each electron according to $j = m_s + m_l$ and the sum of all angular momenta gives the total angular momentum J. However, this coupling scheme has no meaning in practice, since for the ground state the deviation from the Russel-Saunders coupling is only small. Since the ground state is decisive for the magnetic properties of a compound, the L-S coupling scheme can be used to predict the magnetic moment for all elements.

In the following we compare the theoretically determined values for the magnetic moment of 3d and 4f elements with the experimentally determined values μ_{eff}. These are obtained from the molar susceptibility χ using the formula

$$\frac{\mu_{\text{eff}}}{\mu_B} = \sqrt{\frac{3k_B}{\mu_0 N_A}} \frac{\sqrt{\chi T}}{\mu_B} = 2.828 \sqrt{\chi T}$$

Please note that in this equation for the calculation of μ_{eff} the CGS-emu system was assumed. In linguistic usage, the magnetic moment is often given only as μ or μ_{eff} (without μ_B).

In Fig. 10.4, the numerical values for S, L, and J are given for the 3d and 4f elements as a function of the d and f electron numbers, respectively. For an empty and full d and f shells, respectively, all contributions to the magnetic moment are zero, as expected. For a half-full shell ($n(d) = 5$ or $n(f) = 7$), the contribution of the orbital momentum $L = 0$ and the total angular momentum is only dependent on the total spin S. For $n(d) = 4$ or $n(f) = 6$, the coupling of spin and orbital momentum leads to a cancellation of the total angular momentum J.

In Fig. 10.5 the magnetic moments calculated from the total angular momentum are compared with the effectively measured magnetic moments of the 3d and 4f elements. For the lanthanide ions, very good agreement is found between the theoretically expected and measured values. The formula for calculating μ has its justification, experiment and theory agree well. Both the spin and the orbital moment contribute to the total magnetic moment.

A different behavior is observed for the 3d elements. The experimentally determined effective magnetic moment μ_{eff} agrees very poorly with the calculated values for μ. A good agreement is found only for $n = 5$, here the five d-orbitals are all singly occupied and the contribution of the orbital moment to the magnetic moment is 0. In fact, μ_{eff} agrees much better with the values of μ_S. These are the so-called "spin-only" values of the magnetic moment. These values are obtained by assuming that the orbital moment does not contribute to the effective magnetic moment. It is calculated according to the formula

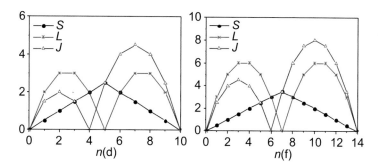

Fig. 10.4 Values for L, S, and J in the ground state for 3d and 4f ions as a function of the d and f electron numbers, respectively

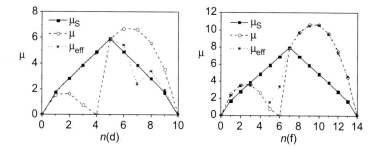

Fig. 10.5 Magnetic moments of 3d ions and trivalent lanthanide ions as a function of the d and f electron numbers, respectively. The measured values μ_{eff} are compared with the calculated "spin-only" values μ_S, in which only the spin moment is taken into account, and the calculated values μ for the total magnetic moment

$$\frac{\mu_S}{\mu_B} = g \cdot \sqrt{S \cdot (S+1)} = \sqrt{n \cdot (n+2)}$$

Here g is the gyromagnetic ratio of the electron (for the free electron $g = 2.0023$), S is the total spin and n is the number of unpaired electrons. Table 10.5 shows the calculated values as a function of the number of unpaired electrons.

Before we address the question why the orbital moment plays only a minor role for the determination of the magnetic moment of 3d elements, we consider *Kossel's displacement theorem*. It states that ions with different nuclear charges but the same number of electrons have the same magnetic moments. This is a fact that we have been using all along. In Table 10.5 we see that this assumption is quite justified.

Quenching of the Angular Orbital Momentum The quenching of the angular orbital momentum for 3d elements is related to the great tendency of these elements to form complexes and the participation of the d orbitals in bonds. In the ligand field, the d orbitals are no longer degenerate but energetically split. This in combination with the weak

Table 10.5 Comparison of calculated (μ_S) and observed (μ_{eff}) magnetic moment as a function of d electron number for 3d elements. EK = electron configuration

Ion	EK	High-spin	n	μ_S	μ_{eff}
Sc^{III}, Ti^{IV}, V^V, Cr^{VI}, Mn^{VII}	$3d^0$		0	0	0
Sc^{II}, Ti^{III}, V^{IV}, Cr^V, Mn^{VI}	$3d^1$	↑	1	1.73	1.6–1.8
Ti^{II}, V^{III}, Cr^{IV}, Mn^V	$3d^2$	↑ ↑	2	2.83	2.7–3.1
V^{II}, Cr^{III}, Mn^{IV}	$3d^3$	↑ ↑ ↑	3	3.87	3.7–4.0
Cr^{II}, Mn^{III}	$3d^4$	↑ ↑ ↑ ↑	4	4.90	4.7–5.0
Mn^{II}, Fe^{III}	$3d^5$	↑ ↑ ↑ ↑ ↑	5	5.92	5.6–6.1
Fe^{II}, Co^{III}	$3d^6$	↑↓ ↑ ↑ ↑ ↑	4	4.90	5.1–5.7
Co^{II}, Ni^{III}	$3d^7$	↑↓ ↑↓ ↑ ↑ ↑	3	3.87	4.3–5.2
Ni^{II}, Cu^{III}	$3d^8$	↑↓ ↑↓ ↑↓ ↑ ↑	2	2.83	2.8–3.3
Cu^{II}	$3d^9$	↑↓ ↑↓ ↑↓ ↑↓ ↑	1	1.73	1.7–2.2
Cu^I, Zn^{II}	$3d^{10}$	↑↓ ↑↓ ↑↓ ↑↓ ↑↓	0	0	0

coupling between spin and orbital momentum leads to the fact that the orbital angular momentum is often quenched. In contrast to this, due to their proximity to the nucleus, the f orbitals are involved in bonding interactions in a much lesser extent, and the energetic splitting is much smaller (insignificant). In order to estimate whether or not a contribution of the angular orbital moment to the effective magnetic moment is to be expected for a certain complex, the following illustrative observation can be used.

For an electron in a particular orbital to have an orbital momentum for a particular axis, it must be able to be transferred by rotation around this axis to an identical degenerate orbital that still has a vacancy for an electron with the appropriate spin [45].

In the free ion, for example, the d_{xy} orbital can be transformed into the $d_{x^2-y^2}$ orbital by rotation around the z axis by 45 °. The same applies to the rotation of d_{xz} by 90 ° along the z axis, we obtain the d_{yz} orbital.

However, if the ion is in an octahedral or tetrahedral ligand field, the d orbitals are split into the t_{2g} and e_g orbitals, which are now no longer degenerate. Part of the orbital moment contribution, e.g. along the z axis (for the pair $d_{xy} \Rightarrow d_{x^2-y^2}$), disappears. Since the degenerate e_g orbitals are not transferable into each other by rotation, they are sometimes called "non-magnetic doublets". For the t_{2g} orbitals, a contribution of orbital angular momentum is still possible in the octahedron for a d^1 or d^2 system. The question whether a complete quenching of the orbital moment is expected can be answered if the coordination geometry and thus the splitting of the d orbitals is known. It must still be taken into account that the quenching of the orbital moment also depends on the size of the splitting of the orbitals. A small splitting leads to a non-complete quenching of the orbital moment, which can be treated as a perturbation in such systems. The orbital moment contribution can lead to significant deviations of the effective magnetic moment from the spin-only value for some complexes.

As an example, we consider an octahedral iron(II) complex in the high-spin state. Here, the two e_g orbitals are each occupied by a single electron and no angular orbital moment contribution to the magnetic moment is to be expected. The three t_{2g} orbitals are occupied by four electrons. The fourth electron (spin-down) can now reside in the d_{xy}, d_{xz} or d_{yz} orbital. The three orbitals can each be transferred into one another by rotation ($xy \rightarrow xz$: rotation about x axis, $xz \rightarrow yz$: rotation about z-axis, $xy \rightarrow yz$: rotation about y axis). Here, an orbital moment contribution to the magnetic moment is expected. Indeed, susceptibility measurements measure an effective magnetic moment in the range of 5.1–5.2, and the spin-only value is 4.9.

10.4   Temperature Dependence of the Magnetic Moment

The temperature dependence of the magnetic moment of an isolated atom or ion follows Curie's law. The reciprocal of the magnetic susceptibility plotted against temperature gives a straight line passing through the origin. Only a few compounds show ideal Curie behavior. Examples would be the temperature dependence of the salts $NH_4Fe^{III}(SO_4)_2 \cdot$ 12 H_2O and $Gd_2^{III}(SO_4)_3 \cdot$ 8 H_2O. In both ions a half-occupied shell is realized (no orbital contribution) and large counterions as well as crystal water ensure that the magnetic centers are well isolated from each other.

Before we look at magnetically concentrated samples, let us look at a special feature of lanthanides using the example of the europium(III) ion. Figure 10.6 shows the magnetic measurement for a europium(III) salt. The europium(III) ion has the electron configuration $[Xe]\ 4f^6$. Thus the total spin is $S = 3$ and the total orbital angular momentum also amounts to $L = 3$. Since the f orbitals are less than half filled, for the ground state the total angular momentum is $J = 0$. The europium(III) ion is diamagnetic in the ground state, even though it has six unpaired electrons. In fact, you can see from the magnetic measurement that at very low temperatures the magnetic moment approaches zero. The next excited states have a total angular momentum of $J \neq 0$ and can be easily occupied by thermal excitation. For

Fig. 10.6 Magnetic behavior of EuCl $_3 \cdot$ 6 H_2O

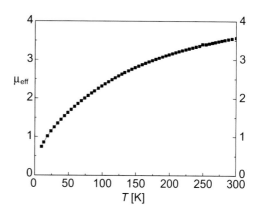

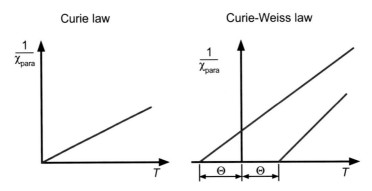

Fig. 10.7 Graphical representation of the Curie law ($\chi \times T = C$) and the Curie-Weiss law $\left(\frac{1}{\chi} = \frac{T-\Theta}{C}\right)$

this reason, a finite magnetic moment is observed at room temperature. From the temperature dependent measurements, the energy difference between the different states can be determined.

In magnetically concentrated samples, various magnetic interactions between the unpaired electrons can be observed. The three most important representatives of these collective interactions are ferromagnetism, antiferromagnetism and ferrimagnetism. Some of these compounds obey the Curie-Weiss law. In contrast to the Curie law, which can be derived mathematically, the Curie-Weiss law is based only on experimental data. The Weiss constant (intersection with the x axis in Fig. 10.7) is zero when no cooperative interactions occur. It is often negative ($\Theta < 0$) for antiferromagnetic interactions and often positive ($\Theta > 0$) for ferromagnetic interactions.

In the following, the characteristics of some important collective phenomena are summarized again. In Table 10.6 the corresponding spin orientations are shown and in Fig. 10.8 the temperature dependence of χ and the product of χT is given.

- Ferromagnetism: Spontaneous parallel orientation of neighbouring magnetic dipoles below the Curie temperature. Above T_C the Curie-Weiss law with positive Weiss constant Θ applies, below T_C the susceptibility is strongly field dependent, often $\chi \propto \frac{1}{H^2}$ applies. Hysteresis loops can be observed (see Fig. 10.23).
- Antiferromagnetism: Spontaneous antiparallel alignment of neighbouring magnetic dipoles below the Néel temperature. Above T_N the Curie-Weiss law with negative Weiss constant applies, below this the susceptibility is weakly field dependent.
- Ferrimagnetism: Spontaneous antiparallel alignment of adjacent magnetic dipoles of different sizes below the Curie temperature, same characteristics as in antiferromagetism.

Table 10.6 Orientation of spins for different collective magnetic phenomena

Spin orientation		Examples
↑↑↑↑↑↑	Ferromagnetic	Fe, Co, Ni, Tb, Dy, Gd, CrO_2
↑↓↑↓↑↓	Antiferromagnetic	MnO, CoO, NiO, FeF_2, MnF_2
↑↓↑↓↑↓	Ferrimagnetic	Ferrites, granates
∧∧∧	Canted antiferromagnetism / spin canting	FeF_3, $FeBO_3$
	Spin fluids, spin spirals and others	Lanthanoides

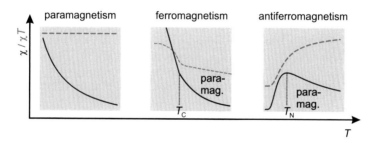

Fig. 10.8 Temperature dependence of χ and χT for a paramagnet, ferromagnet and antiferromagnet. Paramagnetism: χ increases with decreasing T, $\chi T = C$. Ferromagnetism: below Curie temperature T_C drastic increase in χ and χT. Antiferromagnetism: below Nèel temperature T_N drastic decrease of χ and χT

- Special cases:
  - Spin canting, metamagnetism, spin glasses
  - Superparamagnetism: observed in ferromagnetic or ferrimagnetic materials with particle size < size of Weiss domains.

Before we take a closer look at the magnetic exchange interactions and other temperature-dependent phenomena (e.g. spin crossover), we look a little closer at Curie's law. It states that the susceptibility multiplied by the temperature gives a constant. The result is that at high temperatures the susceptibility, and hence the magnetization, is smaller than at low temperatures, and the question arises as to why this is so. To answer this, consider the collection of uncoupled spins given in Fig. 10.9. In the absence of an external magnetic field, the orientation of the magnetic moments is arbitrary and all possible orientations have the same energy. Because of the different orientations, the magnetic moments cancel each other out and the resulting magnetic moment for the entire sample is zero. This changes when the sample is placed in an external magnetic field. The degeneracy of the energy levels is cancelled and a magnetic moment oriented parallel to the external magnetic field is

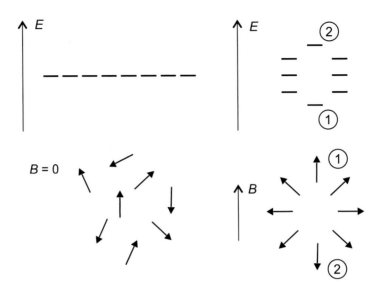

Fig. 10.9 Schematic representation of the orientation of mutually independent magnetic moments in the absence ($B = 0$) and presence ($B \neq 0$) of an external magnetic field

energetically more favorable than an antiparallel oriented magnetic moment. By splitting the energy levels in the presence of an external magnetic field, the states are now occupied following a Boltzmann distribution. This leads to the fact that at very low temperatures only the lowest energy levels are occupied, where all magnetic moments have the same preferred orientation (parallel to the external magnetic field). The magnetization of the sample is large. With an increase in temperature, energetically higher lying energy levels are increasingly occupied. This again leads to a partial cancellation of the magnetic moments and the magnetization becomes smaller. The simple diagram in Fig. 10.9 can also be used to explain the field dependence of the magnetization. At a high external magnetic field, the splitting of energy levels is greater than at a lower external magnetic field. If we keep the temperature constant, this means that with the larger splitting of the energy levels, there is a greater tendency to occupy only the lowest energy levels. At a high external magnetic field, the magnetization becomes larger. At very low temperatures and very high magnetic fields, a state is reached where the magnetization stops increasing. Only the lowest lying energy level is occupied. All spins are aligned parallel to the external magnetic field and the magnetization has reached its maximum. This value is called saturation magnetization.

10.5   Cooperative Magnetism

Ferromagnetism, antiferromagnetism and ferrimagnetism belong to the cooperative magnetic phenomena, which are also called collective magnetism. The basis for these phenomena are spin-spin interactions between neighboring atoms. These interactions lead to the formation of ordered states. The temperature dependence of the magnetic moment for a ferromagnet and an antiferromagnet shown in Fig. 10.8 already indicates that these ordered magnetic states occur at low temperatures and that thermal energy acts against them. Above the Curie or Néel temperature, ferromagnets and antiferromagnets behave like paramagnets. We distinguish the terms ferromagnetic interactions and ferromagnetism (or antiferromagnetic interactions and antiferromagnetism). To observe the material property ferromagnetism, we need 1–3-dimensional spin structures. These often occur in solids. Ferromagnetic interactions can also be observed in a molecule with two spin-bearing centers. However, this does not mean that this molecule behaves like a ferromagnet and shows, for example, hysteresis of magnetization. At this point, it is worth noting the difference between hysteresis of magnetization, as we observe in ferromagnets below the Curie temperature, and thermal hysteresis, as we observe in spin crossover compounds. In iron(II) spin crossover compounds, we observe a temperature-dependent transition between diamagnetism and paramagnetism, which can be accompanied by hysteresis if the intermolecular interactions are sufficiently strong. In the case of ferromagnets, we require at least a 1-dimensional spin structure (coupled spin centers), which should also have some periodicity. Here, an external magnetic field is used to switch the alignment of the coupled spin centers ($\uparrow \uparrow \ \uparrow \uparrow$ or $\downarrow \downarrow \ \downarrow \downarrow$), which is maintained when the external magnetic field is turned off. This periodicity need not coincide with the periodicity of the atoms. An exception to this are the so-called single molecule magnets (SMM). This term applies to molecules that have a large but finite number of coupled metal ions and exhibit hysteresis effects with a molecular origin. These compounds will not be considered further in the following.

In cooperative magnetic phenomena, the coupling between electrons of neighbouring atoms is stronger than the coupling between electrons in one atom. There is an interplay between Hund's (first) rule ($S = $ max) and the Pauli exclusion principle (antiparallel alignment of electrons due to Coulomb interactions). The exchange interactions can occur directly between two metal centers (direct exchange). This occurs, for example, in metals and alloys. The second type is indirect exchange. Here the exchange interactions are mediated by non-magnetic (diamagnetic) bridging atoms. As examples, we consider super-exchange, double-exchange and spin polarization, as well as the concept of magnetic orbitals. To explain the exchange interactions, it does not matter in principle whether one considers the solid state (e.g., oxides or halides) or a multinuclear complex. Examples from both fields are always given in the following. Only in the case of direct exchange between two metal centers is it difficult to find appropriate examples in coordination chemistry and we will consider only the metals. One aspect of particular interest to the coordination chemist is the possibility of controlling the magnetic properties by a clever choice of ligands and metal centers. For simplicity, spin-orbit interactions will be neglected for the reminder of the discussion.

10.5.1 Exchange Interactions

In order to better understand the exchange interactions, we start with the interaction between two metal centers, both carrying the spin $S = \frac{1}{2}$. If there is no interaction between the two centers, then the orientation of both spins with respect to each other is arbitrary. We have two uncoupled metal centers. If an interaction takes place, there are two possible states: the two spins can preferentially align parallel to each other and the total spin of the system is $S = 1$. The second variant involves an antiparallel alignment of the neighboring spins and a total spin of the system of $S = 0$. Depending on which state is the ground state, we have antiferromagnetic (ground state $S = 0$) or ferromagnetic (ground state $S = 1$) interactions between the two metal centers. The energy difference between the two states is the coupling constant J, which is defined as.

$$J = E(S{=}0) - E(S{=}1)$$

The coupling constant can be determined *via* temperature-dependent magnetic measurements, since the occupation of the ground state and the excited states follows a Boltzmann distribution. It is positive for ferromagnetic interactions and negative for antiferromagnetic interactions. In Fig. 10.10 the energy diagram for such a coupled system is given. Upon application of an external magnetic field, the $S = 1$ state splits into three now non-degenerate energy levels.

There are two questions to be answered below:

- What conditions must be fulfilled for exchange interactions to occur at all?
- Under what conditions are ferromagnetic or antiferromagnetic interactions observed?

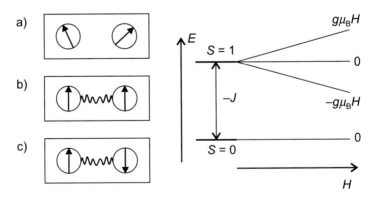

Fig. 10.10 Schematic representation of the interaction between two centers with $S = \frac{1}{2}$. In case (**a**), the orientation of the spins is independent of each other. In case (**b**), a spontaneous parallel position of the spins occurs, a ferromagnetic interaction occurs. In case (**c**), a spontaneous antiparallel position of the spins occurs, an antiferromagnetic interaction occurs. In a coupled system, both possibilities, b and c, can always occur. The difference is which of the two states is the ground state. On the right side the energetic order is given for a system where antiferromagnetic interactions occur

In the further course of the discussion, orbitals with an unpaired electron are called "magnetic orbitals". For any interaction to take place at all, the two orbitals with the unpaired electrons (the "magnetic" orbitals) must be close enough to each other. If the distance is too large, no exchange interaction takes place and Curie behavior is observed as for a paramagnet. For this, we consider the overlap density ρ between the two orbitals. For the occurrence of interaction ρ should be $\neq 0$. The next question is whether overlap between the two magnetic orbitals is possible. This means that even if the two magnetic orbitals are very close, overlap does not necessarily occur. If the overlap integral S is finite, then overlap occurs and antiferromagnetic interactions dominate. One can think of this, similar to a molecular bond, as bonding and antibonding interactions and there is an energetic splitting of the states. This leads to spin pairing in the ground state. If overlap of the orbitals is not possible despite close proximity, then the orbitals are orthogonal to each other and the overlap integral $S = 0$. In this case, both orbitals have similar energy and are occupied according to Hund's rule. We observe ferromagnetic exchange interactions. In general, ferromagnetic interactions are always weaker than antiferromagnetic interactions and when both occur simultaneously in a system, the antiferromagnetic interactions usually dominate. For this purpose, we consider the exchange energy ΔE, which is composed of the potential exchange energy P and the kinetic exchange energy K.

$$\Delta E = P - K$$

The potential exchange energy stands for the parallel orientation of the electron spins. The electrons are in different orbitals, which are occupied according to Hund's rule. The kinetic exchange energy stands for the antiparallel orientation of the spins. The orbitals make a substantial contribution to a bond, they overlap. If the exchange energy (which can be equated with the coupling constant J) is positive, the ferromagnetic interactions dominate; if it is negative, the antiferromagnetic interactions dominate.

10.5.2 Magnetism of Metals

The magnetism of metals is an example of the direct interaction between magnetic orbitals and provides a first insight into the discussion of magnetic interactions. In order to understand the magnetic properties of metals, it is convenient to use the band model to describe the bonding relationships in metals, which is also used to explain conductivity (conductor, semiconductor, insulator).

The band model assumes that the atomic orbitals of the individual atoms in the crystal bond overlap and that molecular orbitals are formed. The more bonds are formed to neighboring atoms, the more molecular orbitals are formed, which are energetically closer and closer together. When the orbitals of enough atoms overlap, an energy band is formed consisting of energy levels that are lined up almost without gaps. For the beginning we start

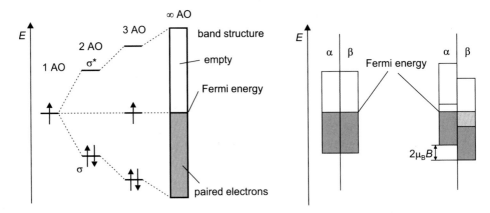

Fig. 10.11 Chemical bond in metals according to the band model. Left: The overlap of atomic orbitals in the crystal results in the formation of energy bands. Right: In the presence of an external magnetic field, an energetic splitting occurs between the electrons with α and β spin. The number of electrons in the conduction band with an orientation of the magnetic moment (we consider only the spin) parallel to the external magnetic field increases at the expense of electrons with oppositely oriented magnetic moment. This distribution is practically independent of the measurement temperature. A weak, temperature-independent paramagnetism is observed

with the lithium atom, where the valence electron is in the 2s orbital. The overlapping of n atomic orbitals leads to the formation of n molecular orbitals, each of which can be occupied by a maximum of two electrons. Since each lithium atom contributes one valence electron to the occupation of the molecular orbitals, half of the energy band is occupied by electrons. The conductivity of metals depends on these partially occupied energy bands. The unoccupied electron states of the energy band can be reversibly filled with electrons that are released elsewhere. This is not possible for fully occupied or empty energy bands. To explain the magnetic properties of metals, we need to look at the width of the energy bands. This depends, among other things, on the size of the metal crystals. In Fig. 10.11 we see on the left-hand side that as the number of atoms in the crystal increases, the bands become wider and wider. A factor of considerable relevance to magnetism is the extent to which the individual orbitals of the metal atoms interact with each other. This in turn depends on the spatial proximity (distance) and the coordination number. A large bandwidth represents (for the same number n of atoms) a significantly larger energetic distance between the individual molecular orbitals than a small bandwidth. The more intense the interaction between the atomic orbitals, the larger the bandwidth of the energy band. This can be equated with the amount of overlap between the atomic orbitals. The Fermi energy (sometimes called the Fermi edge or Fermi limit) indicates the energy up to which the energy band is occupied. If it is exactly half, all spins are paired (occupation according to the Pauli exclusion principle) and the metal is diamagnetic. Nevertheless, a weak paramagnetism is often observed (also in our example lithium), which is approximately independent of temperature. This paramagnetism is due to the conduction electrons (the atomic cores are diamagnetic due to the closed shells) and is much smaller than the temperature-dependent Curie paramagnetism of particles with unpaired electrons. In the presence of an external magnetic field, the energy levels of α and β spin shift relative to each other and an uneven distribution is obtained. The orientation of the magnetic moment of the electrons parallel to the external magnetic field is slightly preferred. This is shown on the right side of Fig. 10.11. When the external magnetic field is switched off, the occupation difference is cancelled again. It is worth pointing out the difference to classical Curie paramagnetism. In Curie paramagnets, there are always unpaired electrons which are randomly oriented in the absence of an external magnetic field and only become oriented in the presence of an external magnetic field. The stronger the external magnetic field, the better the individual spins are oriented. The extent of the alignment is temperature dependent. The higher the temperature, the greater the "disorder" of the individual spins. This means that maximum parallel alignment of the individual spins is only achieved at low temperatures and high magnetic fields.

In the absence of an external magnetic field, the metals are diamagnetic, i.e. all spins are paired. In the presence of an external magnetic field, there is an unequal occupation of α and β spin resulting in a magnetic moment. In contrast to the Curie paramagnet, the individual spins are not independent of each other because they occupy the same energy band. Due to Hund's first rule, all spins are aligned parallel to each other, independent of the ambient temperature. A temperature independent paramagnetism is observed. The extent of magnetization depends only on the external magnetic field.

Next, we ask what happens when the orbitals of the valence electrons overlap less well. A first clue to this is provided by the Bethe-Slater curve. Here, the exchange energy (or coupling constant) of a metal is plotted against the quotient a/r (Fig. 10.12), where a is the atomic distance and r is the radius of the d or f shell. The ratio a/r is, so to speak, a

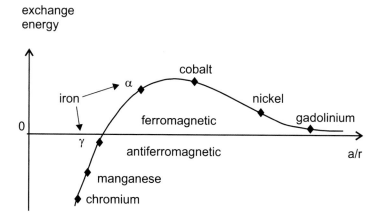

Fig. 10.12 Bethe-Slater curve to explain the magnetism of metals. The ratio a/r is the quotient of the atomic distance a and the radius r of the d or f shell. The plot illustrates the influence of the atomic distance on the exchange interaction

Fig. 10.13 Fermi energy as a
function of the bandwidth ΔE of
the energy bands

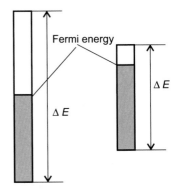

measure of the extent of overlap of the atomic orbitals. We see that with increasing ratios ferromagnetic interactions occur, but these become weaker again when the ratio becomes very large, i.e. the orbitals overlap only very poorly (as in the case of gadolinium).

The ferromagnetism of gadolinium can be explained by the model of *magnetism of localized electrons* [46]. This model assumes that the f electrons of gadolinium are insignificantly involved in bonding because they are highly contracted (very close to the nucleus). Accordingly, the splitting between the bonding and antibonding levels of the resulting molecular orbitals is small and the small splitting results in a small bandwidth ΔE. Because of the small bandwidth, the individual energy levels are very close to each other and, according to Hund's rule, are all initially occupied simply by parallel spin. The Fermi edge lies at the top of the energy band (Fig. 10.13). The concept is comparable to the high-spin and low-spin states in some complexes. The energy loss that results from occupying the entire band with electrons is less than that which would result from electron pairing. We obtain a ferromagnet in which magnetization is observed even in the absence of an external magnetic field. In fact, it is not the energy bands of the f orbitals that are responsible for the conductivity of gadolinium, but the energy bands created by the overlapping 6 s orbitals.

If the valence electrons are in d orbitals and the distance between the metal centers is also shorter, then the bonding between the metal centers is much stronger and the bandwidth increases. If this is the case, then the distance between the individual energy levels is greater and the spin pairing is preferred.

The relationship shown in the Bethe-Slater curve is very obvious at first glance. Starting with gadolinium, a better overlap of the orbitals initially leads to stronger exchange interactions. If the overlap is too strong, the electrons are distributed among the molecular orbitals according to the Pauli principle. However, it must be added that in the case of iron (and also cobalt and nickel) the explanation is no longer sufficient. The difference between the quotient a/r in γ-iron and α-iron is too small to explain the very pronounced differences in magnetism. To explain the ferromagnetism of iron, the energy band must first be divided into the α and β spins, as already shown for lithium in Fig. 10.11 on the right. This representation is also called spin-resolved representation. Calculations show that for a correct description of the electronic structure of α-iron, the α and β levels are energetically shifted with respect to each other even in the absence of an external magnetic field and are therefore occupied differently. A spin polarization takes place. This results in a residual magnetic moment which is responsible for the ferromagnetism. In the non-magnetic α-iron, strongly antibonding Fe-Fe states would be occupied, which is equivalent to a loss of energy for the entire system. There are now two ways to remedy this condition. The first would be a geometric distortion of the lattice, similar to the Jahn-Teller effect discussed for complexes. For Jahn-Teller distorted complexes, a gain in energy for the system is the driving force. In the case of α-iron, we know that the structure is not distorted. The energy gain is achieved *via* a distortion of the electronic structure, the electronic symmetry is degraded by removing the equivalence of α and β spin [47].

10.5.3 Orthogonal Orbitals

The reasons for the occurrence of ferromagnetism discussed for metals are, at first sight, difficult to transfer to complexes (or inorganic solids such as oxides or halides). In the case of complexes, we consider orbitals and not energy bands. The second significant difference is that in complexes the metal centers are usually not directly adjacent to each other, but separated by bridging atoms. A distinction is also made between magnetic insulators (magnetism is mediated by bridging atoms) and magnetic conductors, in which there is a direct interaction between the magnetic orbitals. Two aspects can be transferred from metals to complexes. A certain spatial proximity of the magnetic orbitals (these can also be orbitals delocalized *via* the ligand) is a prerequisite for an interaction to take place. The closer the magnetic orbitals are to each other, the stronger the interaction. The second important point is whether and to what extent the orbitals overlap. Using gadolinium as an example (magnetism of localized electrons), we have seen that ferromagnetic interactions arise because of a weak interaction between the magnetic orbitals. The importance of this point for complexes is best illustrated by the concept of orthogonal magnetic orbitals.

We consider two magnetic orbitals that are spatially close enough to interact with each other. A good example for this is the complex given in Fig. 10.14 on the left. The ligand used here was designed to coordinate two different metal ions in two independent steps. Because of the very close spatial proximity of the two metal ions, we can initially assume a direct interaction between the metal ions and ignore the influence of the bridging ligands (the two oxygen atoms O1 and O2). In order to obtain ferromagnetic interactions in the illustrated complex, the magnetic orbitals (i.e., the orbitals with the unpaired electrons) must be orthogonal to each other. The term orthogonal means that, for reasons of symmetry, no overlap is possible between the orbitals even though they are in close spatial proximity. In Fig. 10.14 on the right, we see from the magnetic measurement that this appears to be the case for the copper(II)-vanadium(IV) complex shown. The product of magnetic susceptibility (χ_M) and temperature increases with decreasing temperature, as we would expect for ferromagnetic interactions. From the same ligand, a corresponding

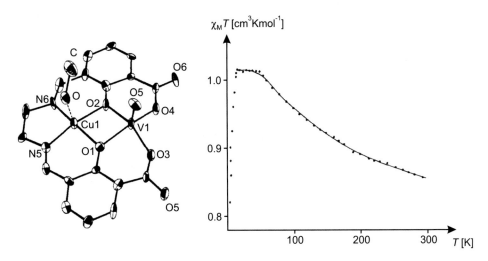

Fig. 10.14 Left: Structure of a dinuclear copper(II)-vanadium(IV) complex. Right: Result of the magnetic measurement on this complex. Shown is the product of $\chi_M T$ as a function of T. On cooling (from *right* to *left*), an increase in the $\chi_M T$ product is observed until a plateau is reached below about 50 K. Adapted from Ref. [48]

binuclear complex with two copper(II) ions can also be prepared. In this complex, not shown here, antiferromagnetic interactions are observed and the product of $\chi_M T$ decreases with decreasing temperature [48].

To understand the different magnetic behavior of the two complexes, we need to determine the magnetic orbitals and consider their relative positions. We see in the structure of the dinuclear complex shown in Fig. 10.14 that the copper(II) ion has an approximately square planar coordination environment. From the electron configuration we determine the number of valence electrons, which in the case of the copper(II) ion is nine 3d electrons. By splitting the d orbitals in a square planar ligand field (see Sect. 4.3.4) we can now determine which of the five d orbitals is occupied by only one electron and is therefore our magnetic orbital (all others are twice occupied). The unpaired electron of copper is located in the $d_{x^2-y^2}$ orbital. The vanadium(IV) ion has only one valence electron, also in the 3d shell. The coordination geometry in this case is square pyramidal, here it has to be taken into account that the oxido ligand (O5 in the structure) is much stronger bond to the vanadium than the other four oxygen atoms because of the double bond character. Therefore, the splitting of the d orbitals in the ligand field corresponds approximately to a compressed octahedron and the lowest energy orbital is the d_{xy} orbital, which accommodates the one unpaired electron [48]. Figure 10.15 shows the relative orientation of the magnetic orbitals for the copper(II)-vanadium(IV) complex and the copper(II)-copper(II) complex.

In the copper(II) vanadium(IV) complex, the orbitals cannot overlap for symmetry reasons. There is strict orthogonality of the two magnetic orbitals due to their symmetry, and the interaction between the two metal centers is ferromagnetic. The two orbitals are

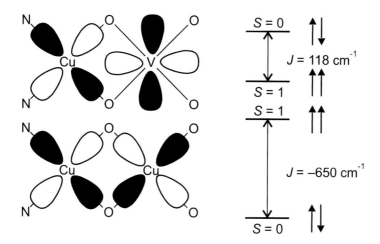

Fig. 10.15 Relative orientation of the magnetic orbitals in a dinuclear copper(II)-vanadium(IV) complex and in the analogous copper(II)-copper(II) complex. In the copper(II) vanadium (IV) complex, the orbitals cannot overlap for symmetry reasons. They are orthogonal to each other and ferromagnetic interactions are observed. In the copper(II)-copper(II) complex, overlap between the two d orbitals is possible and antiferromagnetic interactions are observed. Adapted from Ref. [48]

close enough to influence each other and have similar energy. Consequently, they are occupied according to Hund's first rule (total spin maximum). In the copper(II)-copper(II) complex, overlap between the two d orbitals is possible and antiferromagnetic interactions are observed. The interactions can be illustrated here as an extremely weak bond, in which binding and antibonding "molecular orbitals" are formed. A splitting of the orbitals occurs which leads to the fact that the energetically lowest lying orbitals are occupied first.

A third example is a corresponding mixed Cu-Ni complex. Nickel(II) is octahedrally coordinated and has two unpaired electrons as the d^8 system, which corresponds to two magnetic orbitals. The d_{z^2} orbital is orthogonal to the $d_{x^2-y^2}$ orbital of copper, while antiferromagnetic interactions are expected, and indeed observed, *via* the second orbital ($d_{x^2-y^2}$). At low temperatures, only one magnetic moment is observed corresponding to one unpaired electron [44].

Orthogonal orbitals can also be used in solids to explain ferromagnetic or antiferromagnetic properties. One example are oxides with NaCl structure, such as the pair NiO (antiferromagnetic)/$LiNiO_2$ (ferromagnetic). In this example, the super-exchange mechanism is responsible for the magnetic interactions. This is explained in detail in the following section. First of all, it is important to note that when the Ni-O-Ni angle is 180°, antiferromagnetic interactions occur between the two nickel ions, and when the angle is 90°, ferromagnetic interactions occur. If both interactions occur, then the antiferromagnetic interactions dominate and the material is antiferromagnetic. This is the case with NiO, which crystallizes in a NaCl structure. In Fig. 10.16, a section of the NiO structure is shown on the left. It can be seen that two different Ni-O-Ni angles occur, with

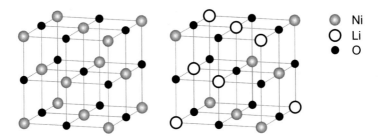

Fig. 10.16 Left: NaCl structure using NiO as an example. Right: Schematic representation of the layer structure of LiNiO$_2$

$90°$ and $180°$. This means that the system only becomes ferromagnetic when there are no $180°$ Ni-O-Ni angles, otherwise antiferromagnetic interactions dominate. This is achieved when a mixed oxide is formed, as is the case with LiNiO$_2$. In this mixed oxide, layers of Li (I) and Ni(III) ions are formed, lining up in such a way that there are no $180°$ angles. The nickel(III) ion is present in the $S = \frac{1}{2}$ low-spin state. The material is now ferromagnetic. Such a structure is shown schematically on the right-hand side of Fig. 10.16.

Super-Exchange Of the examples discussed so far, a direct interaction between the magnetic orbitals takes place. If the two spin-bearing centers are far apart, the overlap density between the two orbitals approaches zero and direct interaction is no longer possible. The magnetic interaction now depends on the orbitals of the bridging ligand. In the following, two different mechanisms will be discussed: super-exchange and spin polarization. The basic idea behind both mechanisms is that the magnetic orbitals are not "pure" d or f orbitals, but also have ligand-based components.

In the super-exchange mechanism [49], the interaction takes place *via* fully occupied s or p orbitals of intermediate diamagnetic bridge atoms. The two electrons in the fully occupied s or p orbitals are aligned antiparallel due to the Pauli principle. The basic idea now is that between adjacent (or overlapping) orbitals, the electrons that are directly adjacent to each other always align in such a way that there is an antiparallel orientation of the spins here as well. Figure 10.17 shows the basic principle for two μ-oxido-bridged metal centers as a function of the M-O-M angle. In each case, the unpaired electron is in a d orbital at the metal center, and from the bridging ligand oxygen, a fully occupied p orbital serves as the bridging orbital. Depending on the M-O-M angle, antiferromagnetic (linear arrangement or obtuse angle, the orbitals can overlap) or ferromagnetic ($90°$ angle, the orbitals are orthogonal to each other) interactions occur. In the case of the $90°$ angle, the d orbitals of the two metal centers each overlap with a p orbital from oxygen. Since the orbitals on the oxygen are two different p orbitals, they cannot overlap. They are orthogonal to each other and ferromagnetic interactions occur. Which of the two exchange mechanisms shown occurs in an angled arrangement depends not only on the M-O-M angle, but also on the magnetic orbital. In the NiO/LiNiO$_2$ example discussed earlier, the

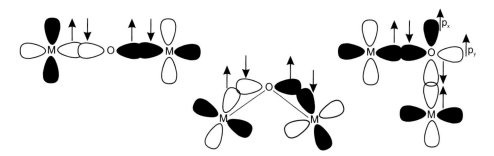

Fig. 10.17 Super-exchange mechanism for a μ-oxido complex with different M-O-M angles. At the 180 $^\circ$ angle and obtuse angles, super-exchange can occur and antiferromagnetic interactions occur. If a 90 $^\circ$ angle is present, then the orbitals at oxygen cannot overlap due to symmetry reasons. The orbitals are orthogonal to each other and ferromagnetic interactions occur

magnetic orbitals from the nickel(II) or nickel(III) ion (octahedral coordination environment, d^8 or d^7) are the $d_{x^2-y^2}$ orbital and/or the d_{z^2} orbital. For both, the orbitals lie on the axes of the metal-ligand bond. With a 90 $^\circ$ angle, only the variant of overlap with the orbitals of oxygen shown in the far right of Fig. 10.17 is possible. If the unpaired electrons are present in the d_{xz}, d_{yz} or d_{xy} orbitals, then the variant shown in the middle would also be conceivable, which would then again lead to antiferromagnetic interactions.

An example of antiferromagnetic interactions due to the super-exchange mechanism is provided by copper acetate. The complex exists as a dimer with the exact composition [(Cu (ac)$_2$(H$_2$O)$_2$], in which one water coordinates as an additional ligand per copper ion. The structure of the dinuclear complex, the result of the magnetic measurements and the exchange mechanism are given in Fig. 10.18. The magnetic measurements clearly show that antiferromagnetic interactions exist between the two copper ions. The $\chi_M T$ product drops at low temperatures, and below 50 K the compound is diamagnetic. As for the copper (II) complexes discussed earlier, the magnetic orbital is again the $d_{x^2-y^2}$ orbital, which is good at forming σ bonds with the orbitals of the bridging acetate ligands. As shown in Fig. 10.18 on the right, super-exchange mediated by the acetate ligands leads to antiferromagnetic interactions [44].

Spin Polarization   In some multinuclear complexes, the magnetic orbitals may not overlap with the σ, but with the π orbital of the ligand. In these cases, the magnetic properties can be explained by the spin polarization mechanism. The spin polarization mechanism is derived from a molecular orbital model proposed by Longuet-Higgins [50] for aromatic hydrocarbons. The starting point was the question of when organic diradicals are stable and the relevance of valence structural formulas in answering this question. Calculations have shown that ferromagnetic interactions between two radicals are possible if they are linked by an *m*-phenylene bridge, since the two unpaired electrons are in a pair of degenerate SOMOs (SOMO = single occupied molecular orbital) of the same orthogonality. Again, the first of Hund's rules finds its application. Interestingly, as shown in Fig. 10.19, no

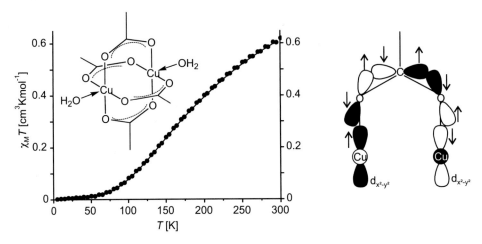

Fig. 10.18 Structure of copper acetate and result of magnetic measurement. The antiferromagnetic interactions are due to super-exchange mediated via the bridging acetate ligands

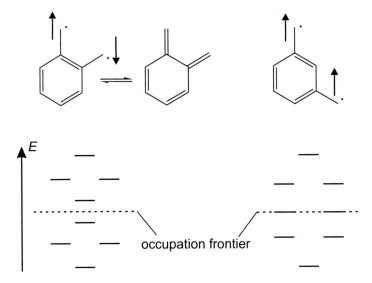

Fig. 10.19 Section of the MO scheme and valence structural formulas of two organic diradicals. For the *ortho*-bridged diradical shown on the left, an alternative valence structural formula can be constructed in which the diradical character is lost. The diamagnetic ground state is consistent with the calculated MO scheme. For a *meta*-bridged diradical, this possibility does not exist and the MO scheme is also consistent with a paramagnetic ground state [50]

alternative valence structural formula can be established for *meta*-bridging, where the diradical character is lost. One result of this behavior is the alternating arrangement of α and β spins in the bridging atoms, called spin polarization. In other words, the spin

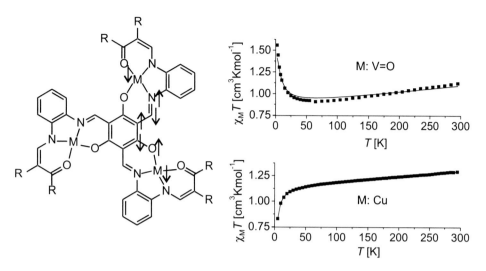

Fig. 10.20 Comparison of the magnetic properties of a trinuclear *meta*-bridged copper(II) complex and a trinuclear *meta*-bridged vanadyl(IV) complex. The arrows in the structural formula indicate the spin polarization mechanism

densities of neighboring atoms in a π conjugated system prefer opposite signs. We have, so to speak, an antiferromagnetic interaction of electrons in non-orthogonal 2p orbitals. This mechanism is very suitable for making predictions about the existence of organic polyradicals, but can also be applied to complexes when the exchange interactions take place via the π system of a planar, sp^2-hybridized ligand.

We look at two examples of complexes. In both cases, ligands were prepared that can be used to make targeted *meta*-bridged trinuclear complexes. Of the ligand shown in Fig. 10.20, a trinuclear copper(II) complex and a trinuclear vanadium(IV) complex were prepared. In the copper(II) complex, the $d_{x^2-y^2}$ orbital is again the magnetic orbital. The ligand is designed to give a square planar ligand field. The $d_{x^2-y^2}$ orbital can only overlap with the σ bonds of the ligand. Under these conditions, only super-exchange can be considered as the exchange mechanism and antiferromagnetic interactions dominate. Accordingly, a decrease in the product of $\chi_M T$ at low temperatures is observed in the magnetic measurement. In the vanadyl(IV) complex, which has a square pyramidal coordination environment as in the last example, the magnetic orbital is the d_{xz} or d_{yz} orbital. This is because the equatorial ligand used here is much stronger and the splitting of the d orbitals in the ligand field this time corresponds to that of a stretched octahedron. Both d orbitals can overlap with the π-system of the planar ligand and the spin polarization leads to ferromagnetic interactions, as shown in Fig. 10.20. Because of the large distance between the metal centers, the interactions are much weaker than in the examples discussed so far. Figure 10.21 shows the magnetic properties of a very similar copper(II) complex, but here

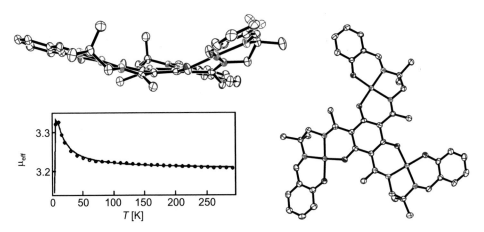

Fig. 10.21 Magnetic properties of a saddle-shaped, trinuclear *meta*-bridged copper(II) complex. (Adapted from Ref. [51])

ferromagnetic interactions are observed. Depending on the structure of the ligand, there are different possibilities for the transfer of spin polarization to the π system of the ligand [51].

Double Exchange The double-exchange mechanism [52] can only be observed in mixed-valent complexes or solids. An example would be mixed-valent iron(II/III) complexes (or solids) in which the coordination environment of the two iron centers is nearly identical. In this case, the oxidation states cannot be unambiguously assigned to the individual atoms; the additional electrons are delocalized, so to speak, over several metal ions. This (thermally activated) hopping of the electrons can only take place if the spin of the electron does not change in the process. Only then is the hopping electron in an energetically favorable state according to Hund's first rule. This results in a parallel aligned spin for neighboring metal ions, which corresponds to a ferromagnetic interaction. This mechanism is always accompanied by a high electrical conductivity of the material.

A good example of double exchange is perovskites of composition $La_{1-x}Pb_xMnO_3$. If $x = 0$, then manganese has oxidation state +3, and if $x = 1$, the oxidation state is +4. Both situations order antiferromagnetically. For $0.2 \le x \le 0.8$, a mixed valence compound is obtained. Ferromagnetic interactions are observed, accompanied by an increase in conductivity.

Spin canting In all the examples discussed so far, the magnetic moments of the metal centers are oriented parallel or antiparallel to each other. However, this need not always be the case. There are examples where an angle between the spin orientations occurs, the canted antiferromagnetism has already been listed in Table 10.6. Often such behavior is related to external parameters (packing in the solid, or, as shown in Fig. 10.22, due to the

Fig. 10.22 Example of an iron
(II) coordination polymer in
which the iron centres are each
rotated 90 ° relative to one
another with the result of the
magnetic measurement

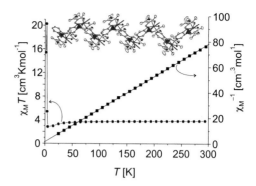

coordination environment). In the iron(II) coordination polymer shown in the figure, the
iron centers are twisted approximately 90 ° to each other. Antiferromagnetic interactions
occur across the bridging ligands. Due to the special orientation of the iron centers and thus
the magnetic moments, the magnetic moment does not go down to zero at low
temperatures, but we observe a spontaneous magnetization. Due to the relatively weak
interactions (the ligand is quite large), this only occurs at very low temperatures.

10.5.4 Microstructure of Ferromagnets

The occurrence of ferromagnetic interactions does not necessarily mean that the material is
a ferromagnet. The same is true for ferrimagnetic interactions. For the occurrence of
ferromagnetism or ferrimagnetism we need, in addition to the right interactions, at least a
one-dimensional extended structure of coupled centers. In most cases, these are 2D or 3D
structures. This is reflected by the fact that most ferro- or ferrimagnets are solids.

   A special property of the ferromagnet is the hysteresis of the magnetization. In a
ferromagnetic material, due to the strong but short-range exchange interactions, domains
are formed in which all centers have the same spin orientation. These domains are called
Weiss domains and have a size in the range of 1–100 μm. In Fig. 10.23, the Weiss domains
are shown schematically. For a non-magnetized material, the orientation of the magnetic
moment of the individual domains is disordered, so that no magnetization is observed for
the entire material (Fig. 10.23 left). When an external magnetic field is applied, the
magnetic moment in all domains is aligned parallel to the external magnetic field
(Fig. 10.23 middle). Long-range interactions (dipole interactions) now ensure that after
the external magnetic field ($H = 0$) is switched off, the magnetization (or magnetic
induction) does not disappear, but only decreases to so-called remanence (B_m). The
magnetization disappears only when the coercivity field strength (H_C) is applied in the
opposite direction. Here we distinguish between soft magnetic and hard magnetic
materials. The two materials differ in the coercivity, which is a measure of the width of
the hysteresis. In soft magnetic materials, which are used in transformers and motors,

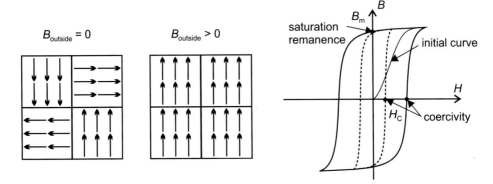

Fig. 10.23 Schematic representation of the Weiss domains for a non-magnetised ferromagnet *(left)* and a ferromagnet magnetised with the aid of an external magnetic field *(middle)*. The magnetization curve of a ferromagnet is shown on the right. The new curve for the first magnetization of the material and the hysteresis curves for a soft magnetic *(dashed line)* and a hard magnetic *(solid line)* material are shown

among other things, the coercivity is small ($H_C < 10$ A/cm). Only a small external magnetic field is needed to reverse the magnetization of the sample. We are familiar with hard magnetic materials in the form of permanent magnets, for example. Here, a large coercive field strength is required to reverse the magnetization of the material.

In this context, the phenomenon of superparamagnetism should be briefly mentioned. This property occurs with very small particles of a ferro- or ferrimagnetic material, where the particle size is smaller than the domain size of the Weiss domains. For superparamagnetic materials, the hysteresis of the magnetization disappears. The reversal of the external magnetic field causes the particle to simply flip over. Hysteresis can be observed when particles are fixed in a matrix such that movement is no longer possible.

10.6   Spin Crossover

For 3d elements, in addition to the collective magnetism phenomena already mentioned, another phenomenon can be observed in temperature-dependent measurements of magnetic susceptibility, for example a thermally induced spin transition known as *spin crossover*.

Spin crossover is a fascinating example of bistability in coordination compounds. This phenomenon is most frequently observed in octahedral complexes. Two different spin states are conceivable for a metal center with a d^n electron number of $n = 4$–7. At first, we consider iron(II) complexes with a d^6 electron configuration. For the hexaaqua complex with water as the weak-field ligand, the energetic splitting between the t_{2g} orbitals and the e_g orbitals (Δ_O) is much smaller than the spin-pairing energy P that has to be applied when one orbital is occupied by two electrons ($\Delta_O \ll P$). The electrons are distributed among the

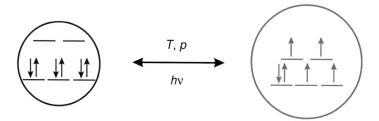

Fig. 10.24 Schematic representation of a spin crossover for a compound with d^6 electron configuration. The transition between the LS state *(maximum number of paired electrons, left)* to the HS state *(maximum number of unpaired electrons, right)* can be triggered by various external parameters such as change of temperature, change of pressure or exposure to electromagnetic radiation. The coordination number of the compound (in this case 6) remains constant

orbitals according to Hund's rule in such a way that all are initially singly occupied. A high-spin (HS) complex is obtained, in which the maximum possible number of unpaired electrons is present. With the strong-field ligand cyanide, the splitting between the t_{2g} orbitals and the e_g orbitals is much larger than for the aqua complex and also much larger than the spin-pairing energy ($\Delta_O \gg P$). Accordingly, it is energetically more favorable that initially only the lower-lying t_{2g} orbitals are occupied. A low-spin (LS) complex is obtained in which the minimum number of unpaired electrons is present. For some ligands, it happens that neither condition is clearly satisfied ($\Delta_O \approx P$). This condition is observed, for example, in the iron complex $[Fe(ptz)_6](BF_4)_2$ (ptz = 1-propyltetrazole). In such compounds, a spin transition can be observed, triggered by a change in external parameters (temperature, pressure, irradiation of light, …) and associated with a change in the physical properties of the compound. This phenomenon is called spin crossover (SCO) (Fig. 10.24. [53, 54]).

When talking about the phenomenon of spin crossover, or to be more precice spin transition, one has to be aware of the fact that there are two different forms of this event – spin transtition with and without change of the coordination number. Colloquially, one usually means the latter and also in this section we restrict ourselves to the variant of the spin crossover with preservation of the coordination number. For the sake of completeness, however, an example of a compound class with spin transition while changing the coordination number is given. These are nickel(II) complexes $[Ni(L)_2]$ where L is a ketoenolate (Fig. 10.25). If the substituents R are bulky, the molecule exists as a monomer at room temperature as shown in Fig. 10.25 on the left. The compound is red and diamagnetic (square planar ligand field with d^8 configuration). At lower temperatures or smaller substituents R, the compound changes to a green, trimeric paramagnetic species in which the coordination number increases from four to six (octahedral).

The effect of spin crossover while preserving the coordination number was discovered in 1931 by *Cambi et al.* on iron(III) tris(dithiocarbamates) (Fig. 10.26) [55]. The first iron (II) complex was synthesized and studied in 1964 [56] and from then on a lively exploration of this phenomenon began. Highlights in spin crossover research were the discovery of

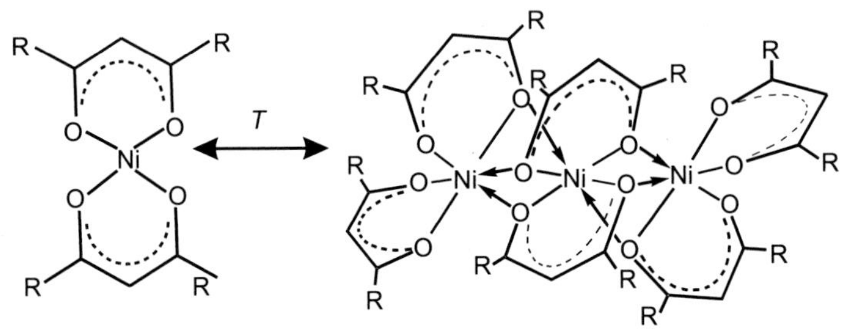

Fig. 10.25 Schematic representation of a spin transtition with a change in coordination number. In the monomeric compound *(left)*, the d^8 nickel(II) ion is present in a square planar coordination environment. The compound is red and diamagnetic. At lower temperatures or small substituents R, a trimeric modification is formed. The coordination number changes from four to six and the compound is now paramagnetic

Fig. 10.26 General formula of iron(III) tris(dithiocarbamate) complexes

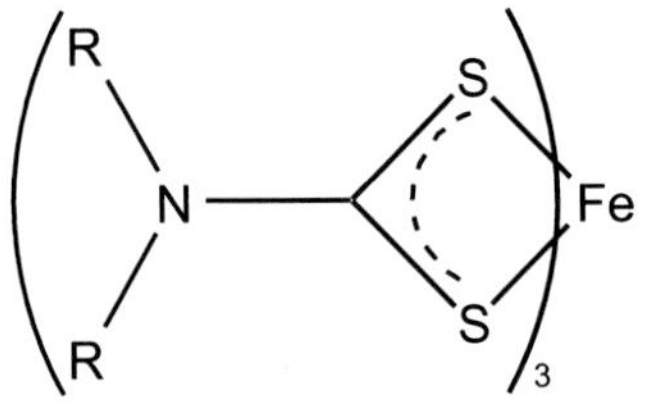

the LIESST effect (switching of the spin state with light, 1984 [57, 58]) and the realization of nanostructuring and functionalization of these compounds, which is of particular interest for potential applications. The spin crossover phenomenon is particularly common in iron (II/III) and cobalt(II) complexes, but compounds with thermal spin transition are also known from nickel(II), cobalt(III), manganese(III) and chromium(II) [53, 54].

10.6.1 Theoretical Considerations

The equation $\Delta \cong P$ vividly represents the condition for a thermal spin transition, but it is very inaccurate. This is due to the fact that the ligand field splitting Δ (or the parameter $10\,Dq$) in the LS state is different from that in the HS state, and also the spin pairing energy of the two states is not necessarily equivalent. The Δ used in the equation does not belong to either the HS or LS state, but represents the intersection of the potential curves of the HS and LS states. Therefore, for a correct consideration of the spin crossover phenomenon, this approximation should be avoided. At the molecular level, the spin transition corresponds to an "intraionic electron transfer", i.e. the electrons remain in the immediate vicinity of the metal ion. Since in the HS state the antibonding e_g orbitals are occupied with more electrons than in the LS state, the average metal-ligand bond lengths increase for the LS $\rightarrow$ HS transition. The magnitudes are 0.14–0.24 Å for iron(II) compounds, 0.11–0.15 Å for

Table 10.7 Magnitude of ligand field splitting $10 \, Dq$ or Δ_O for iron(II) HS, LS and SCO complexes

$10 \, Dq^{HS}$	$< 11000 \, \text{cm}^{-1}$	HS complex
$10 \, Dq^{HS}$	$\approx 11{,}500\text{--}12{,}500 \, \text{cm}^{-1}$ and	
$10 \, Dq^{LS}$	$\approx 19{,}000\text{--}21{,}000 \, \text{cm}^{-1}$	spin crossover complex
$10 \, Dq^{LS}$	$> 21{,}500 \, \text{cm}^{-1}$	LS complex

iron(III) compounds and 0.09–0.11 Å for cobalt(II) compounds [53]. This increase in bond lengths is the reason for the different splitting of the d orbitals in the ligand field. The distance dependence of the parameter $10 \, Dq$ is given by the following equation:

$$10 \, Dq(r) = 10 \, Dq(r_0) \left(\frac{r_0}{r} \right)^6$$

In order to consider a spin crossover correctly with ligand field theory, the Tanabe-Sugano diagrams are used (see Sect. 5.4). For complexes with weak-field ligands, the HS $^5T_2$ state is the ground state, while above a critical ligand field strength Δ_{crit} the LS $^1A_1$ state is the ground state. Now, only the distance dependence of the d orbital splitting ($10 \, Dq$) has to be taken into account to estimate the change of the ligand field strength as a function of distance. For spin crossover compounds, $10 \, Dq(\text{HS}) < \Delta_{crit} < 10 \, Dq(\text{LS})$ holds. The term crossing point Δ_{crit} in the Tanabe-Sugano diagram can be equated with the crossing point of the two potential wells for the HS and LS states. This picture allows a correct formulation to describe the conditions for a spin crossover: $\Delta E_0 \, (\text{HL}) \approx k_B T$. Here ΔE_0 (HL) corresponds to the difference of the zero-point energies of the potential curves of the HS and LS states. The range of $10 \, Dq$ in which iron(II) spin crossover complexes can be expected is given in Table 10.7 [53].

Thermodynamic Considerations Now that the conditions for the occurrence of a spin crossover have been considered in more detail, the question of the temperature dependence arises next. To understand this, the Gibbs-Helmholtz equation is used (we assume that the pressure remains constant):

$$\Delta G = \Delta H - T \Delta S$$

The ground state in a spin crossover system is in each case the one with the lowest free energy (G), which is composed of an entropy (S) and an enthalpy (H) component. In each case, the Δ refers to the difference between the HS and LS states. We now define the critical temperature $(T_C$ or also $T_{\frac{1}{2}})$ at which the same number of molecules are present in the HS and LS states. It is defined as:

$$\Delta G = 0$$

with:

$$T_C = \frac{\Delta H}{\Delta S}$$

The entropy change is composed of an electronic and a vibrational component. In a perfectly octahedral iron(II) complex, the LS state is single degenerate, while for the HS state there is a 15-fold degeneracy (in terms of total spin and total orbital angular momentum, $((2S + 1) \cdot (2L + 1))$). ΔS_{el} is thus:

$$\Delta S_{el} = Nk_B \ln\left(\frac{\Omega_H S}{\Omega_L S}\right) = 1.882 \, \text{cm}^{-1} \text{K}^{-1} = 22.5 \, \text{JK}^{-1} \text{mol}^{-1}$$

Just to remind you: $1 \, \text{cm}^{-1} = 1.986 \times 10^{-23} \, \text{J} = 11.98 \, \text{J mol}^{-1} = 0.124 \, \text{meV}$.

The vibrational disorder in the HS state is also higher than in the LS state because the metal-ligand bond lengths are longer. Since ΔS is positive, ΔH must also be positive in order to observe a spin transition (in a physically reasonable temperature range). Accordingly, the minimum of the LS potential energy curve must be slightly lower than that of the HS potential energy curve. Figure 10.27 shows the relative positions of the HS and LS potential wells. At low temperatures the enthalpy factor dominates and the LS state is more stable, while at higher temperatures the entropy dominates.

Using a little mathematics, we can now derive an expression to describe the temperature-dependent occupation of the HS state. For this we need another expression for the determination of the free energy:

$$\Delta G = -k_B T \ln(K)$$

K in this case is the equilibrium constant for the equilibrium between the HS and LS states. The following expression can be used for it:

$$K = \frac{\gamma_{HS}}{\gamma_{LS}} = \frac{\gamma_{HS}}{1 - \gamma_{HS}}$$

γ_{HS} and γ_{LS} represent the molar fraction of HS and LS molecules, respectively. The value ranges from 1 (all HS or LS) to 0 (all LS or HS). Since there are only two species in equilibrium, $(1-\gamma_{HS})$ can also be used instead of γ_{LS}. Now we put both expressions together and get:

$$\Delta H - T\Delta S = -k_B T \ln\left(\frac{\gamma_{HS}}{1 - \gamma_{HS}}\right)$$

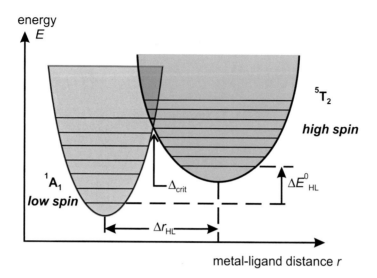

Fig. 10.27 Relative location of the potential wells of the $^5T_2$ high-spin and $^1A_1$ low-spin states for an iron(II) spin crossover complex. Due to the occupation of the antibonding e_g orbitals in the high-spin state, the bond lengths increase during the transition from LS to HS

Transformed, this gives the following expression for the temperature dependence of γ_{HS}:

$$\gamma_{HS} = \frac{1}{1 + e^{\left(\frac{\Delta H}{k_B T} - \frac{\Delta S}{k_B}\right)}}$$

In the case of iron(II) spin crossover compounds, the values of ΔH are in the range of 6–15 kJ/mol and ΔS are in the range of 40–65 JK^{-1} mol^{-1} with spin transition temperatures around 130 K, respectively. About 30% of the measured entropy gain is accounted for by the electronic (magnetic) part. Most of the remaining 70% is distributed (about half – half) to intramolecular stretching and deformation vibrations. Changes in intermolecular vibrations cause only a relatively small proportion [53].

10.6.2 Pressure Dependence

Due to the fact that the complex molecules in the HS state are larger than in the LS state, it is expected that the LS state is stabilized when the pressure is increased. This is generally observed in pressure experiments on iron(II) complexes in the solid state (shift of the spin transition to higher temperatures) [59, 60]. Negative pressure can also be induced when complex molecules are embedded in an appropriate host lattice. For lattices with larger ions (e.g. Zn^{2+}), the HS state is stabilized. The spin transition curve shifts to lower temperatures

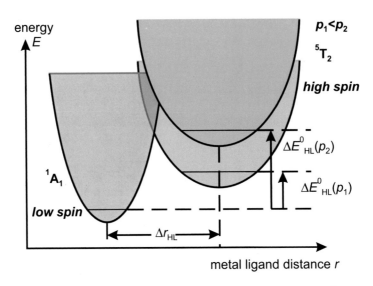

Fig. 10.28 Relative displacement of the potential wells of the T_2 high-spin and $^1A_1$ low-spin states with pressure increase

with increasing dilution and simultaneously becomes more gradual as the cooperative interactions between the individual spin crossover complex molecules are suppressed. Similar observations have been made for Mn(II) and Co(II) host lattices [59]. Figure 10.28 shows the effect of pressure change on the relative position of the HS and LS potential wells.

10.6.3 Switching with Light: The LIESST Effect

The influence of electromagnetic radiation can induce switching between the LS and HS states at low temperatures in some spin crossover compounds. The best known of these phenomena is the light-induced spin transition called Light Induced Excited Spin State Trapping, or LIESST [53]. The prerequisite for the LIESST (LS → HS) and reverse LIESST (HS → LS) effects is two minima in the potential energy curve. In spin crossover systems this requirement is met, although, as discussed earlier, the LS minimum is slightly lower than the HS minimum. This phenomenon was first discovered by *Decurtins et al.* on [Fe(ptz)$_6$](BF$_4$)$_2$ (1984, ptz = 1-propyltetrazole) [57, 58]. The complex is colorless in the HS state and dark red in the LS state. In the single crystal absorption spectrum, bands are observed at 820 nm and 514.5 nm, which can be assigned to the $^5T_2 \rightarrow {}^5E$ and the $^1A_1 \rightarrow$ $^1T_1$ transitions in the Tanabe-Sugano diagram, respectively. The details of this have already been discussed in Sect. 5.4. The relative intensity of these bands can be used to follow the transition curve for the thermal spin transition of this complex, which agrees well with that obtained from susceptibility measurements. If the crystal is irradiated below a temperature of 50 K (the compound is then in the LS state and red) with a wavelength of 514.5 nm ($^1A_1 \rightarrow {}^1T_1$ band), it bleaches out within a very short time and the absorption spectrum typical for the HS state appears. This HS state is metastable with almost unlimited lifetime and only at temperatures significantly above 50 K a relaxation back to the LS state does set in. The temperature at which the return to the LS state occurs is called the T(LIESST) temperature. If the colorless crystal in the HS state is now irradiated at 820 nm ($^5T_2 \rightarrow {}^5E$ band), it regains its red color after a short time. However, the reconversion is not complete ($\gamma_{HS} = 0.1$).

Figure 10.29 shows the mechanism schematically. The mechanism of the LIESST effect was elucidated in detail on the compound $[Fe(ptz)_6](BF_4)_2$ [61]. Upon irradiation of the

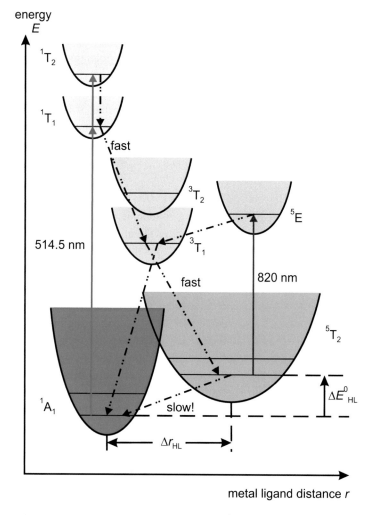

Fig. 10.29 Relative location of the potential wells of the $^5T_2$ HS and $^1A_1$ LS states, as well as the excited states for an iron(II) spin crossover complex with the transitions relevant to the LIESST effect

complex in the low-spin state with green light (514.5 nm), a spin-allowed ($\Delta S = 0$) excitation of the system to the $^1T_1$ or $^1T_2$ state occurs. Two successive fast intersystem crossing steps (spin forbidden, $\Delta S \neq 0$) *via* the intermediate state $^3T_1$ cause a non-radiative relaxation into the HS state $^5T_2$ and the LS state $^1A_1$. Upon continuous irradiation with green light, the system is continuously transferred from the LS state to the HS state until the $^1A_1$ state is emptied and no further excitation can take place. Irradiation of the now HS state with red light (820 nm) results in spin-allowed ($\Delta S = 0$) excitation to the ^5E state. By two fast intersystem crossing steps, again *via* the intermediate $^3T_1$ state, the system relaxes back to the low-spin $^1A_1$ state and the HS $^5T_2$ state. Continuous irradiation with red light leads to the almost complete transition of the complex to the LS state. Direct relaxation $^5T_2 \rightarrow {}^1A_1$ occurs very slowly at low temperatures, so the faster intersystem crossing relaxation pathways are preferred. The metastable light-induced HS state is therefore very long-lived at low temperatures.

Since the discovery of the effect, it has been demonstrated for many iron(II) SCO complexes. This was achieved not only for concentrated compounds but also for dilute solid solutions or SCO complexes embedded in polymer films [53]. The lifetime of the metastable HS state is between 10 and 10^5 s at low temperatures. During relaxation, strong cooperative interactions are noticeable in the concentrated solid state. Systematic investigations are currently underway to determine which factors influence the lifetime of the metastable HS state. An important correlation is the "Inverse Energy Gap Law" of *Hauser et al.* [62]. Here, an inverse proportional relationship between the thermal spin transition temperature $T_{\frac{1}{2}}$ and T(LIESST) is predicted. This relationship is confirmed by *Létard et al.* who performed systematic studies on a large number of SCO complexes. Here, yet another factor is held responsible for the magnitude of T(LIESST) – the rigidity of the inner coordination sphere around the SCO center. The more rigid the coordination sphere around the iron center is, the higher T(LIESST) is for the same $T_{\frac{1}{2}}$ [63].

The highest T(LIESST) temperature to date was determined for an iron(II) complex in which the iron center in the HS state has coordination number 7. During the thermal transition to the LS state, one of the coordinative bonds is broken, only to be rejoined in the light-induced HS state. This bond linkage presumably leads to the high stability of the light-induced HS state [64]. This stabilization can go so far that a light-induced HS state is observed at low temperatures even for a pure LS complex (studied up to 400 K) [64]. Of particular interest is the possibility of switching back and forth between the HS and LS states within a hysteresis around room temperature using light (laser pulse). This was first realized for the complex [Fe(pyrazine) Pt(CN)$_4$] [65]. The metastable HS state can be excited not only optically but also by a number of other mechanisms. One possibility was observed in Mössbauer emission spectra where the HS state was produced by ^{57}Co (EC) ^{57}Fe nuclear decay (EC = electron capture). This nuclear decay induced phenomenon has been named NIESST (Nuclear decay Induced Excited Spin State Trapping) by analogy with LIESST [53]. Hard or soft X-rays can also lead to the occupation of the metastable HS state (HAXIESST = HArd X-ray Induced Excited Spin State Trapping and SOXIESST = SOft X-ray Induced Excited Spin State Trapping) [66].

10.6.4 Cooperative Interactions and Hysteresis

If one examines a spin transition in solution, a gradual spin crossover is always observed, which follows a Boltzmann distribution. The corresponding mathematical expression for the temperature dependence has already been derived. In the solid state, different types of spin crossover behavior can be observed. It can be complete or incomplete, gradual or abrupt, and steps or hysteresis can occur (Fig. 10.30) [67]. The prerequisite for the occurrence of thermal hysteresis loops are cooperative interactions between spin crossover centers. One assumption is that the information of the volume change of the single molecule during the spin transition is propagated from a starting point in the crystal from one molecule to the next. In this context, one speaks of elastic interactions between the complex molecules. Various models have been developed for the mathematical description of these interactions. Mentioned here are the model of elastic interactions [67], the model of internal pressure, the Ising model or the domain model. All of them lead to similar mathematical expressions that can be converted into each other [68]. The basic idea is

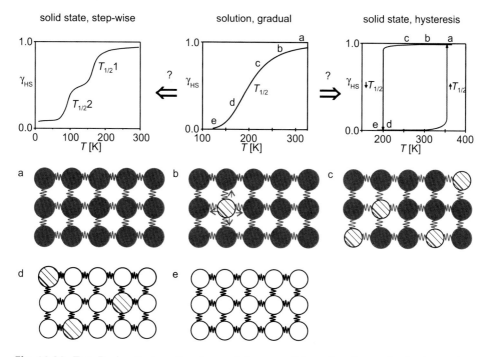

Fig. 10.30 Top: During the transition from solutions to solids, various intermolecular interactions can lead to changes in spin crossover behaviour: *left:* stepwise spin transition, *middle:* gradual spin transition, *right:* spin transition with hysteresis. Bottom: schematic representation of the concept of the regular solution model, which is often used to explain hysteresis. The "springs" illustrate the intermolecular interactions, white are the molecules in the LS state, black in the HS state and hatched stands for molecules that adapt their spin state to the system due to the intermolecular interactions

that, starting from the HS state (Fig. 10.30, point a), when the temperature decreases, said interactions first prevent the spin transition at individual metal centers (Fig. 10.30, point b, c), until the temperature is low enough for the system to completely change to the LS state (Fig. 10.30, point d). The same is true for an increase in temperature. Thus, for sufficiently strong interactions, thermal hysteresis loops can occur. Figure 10.30 shows a schematic sketch of the process.

The preparative chemist now has two factors that he can influence: (i) the nature of the interactions and (ii) the dimension of the network. Possible interactions currently discussed in the literature are (i) van der Waals interactions (contacts shorter than the sum of van der Waals radii), (ii) π-π interactions, (iii) hydrogen bonds, and (iv) so-called covalent cross-links. The last variant is particularly appealing to the preparative chemist, as it allows the best control over the dimension of the crosslinking. Systematic studies on a variety of one-dimensional coordination polymers indicate that the covalent bridges in these compounds contribute to the cooperative interactions only when they are rigid and occur in combination with additional interactions between the polymer strands [69, 70]. This experimental finding was also confirmed by simulations on such compounds [71]. The broadest hystereses in structurally elucidated compounds (a crystal structure is essential to accurately investigate the causes of the occurrence of broad hystereses) have so far been observed on complexes cross-linked via hydrogen bonds (broadest hysteresis around 70 K) [72] or where strong π -π interactions are observed (broadest hystereses around 40 K) [73, 74]. In Fig. 10.31, the molecular structure and the result of the magnetic measurements of the compound with 70 K broad hysteresis are given as an example.

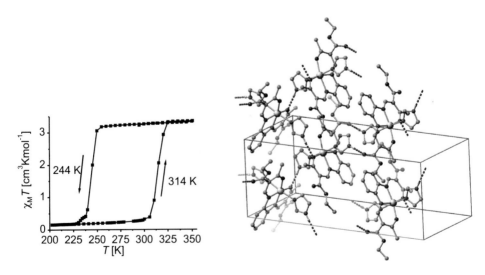

Fig. 10.31 Section of the molecular structure of the compound [FeL$_{eq}$(HIm)$_2$], L$_{eq}$ is a tetradentate Schiff base ligand, and result of magnetic measurement. The hydrogen bonds are shown as dashed lines. Adapted from Ref [72]

10.7   Questions

1. (a) What do the symbols H, B, χ and μ_r mean in magnetochemistry? (b) What is their relationship?
2. Which quantum numbers have an influence on the magnetic moment? What is the connection?
3. (a) When do we speak of cooperative phenomena or cooperative magnetism? (b) Which different cooperative phenomena can occur? Sketch the course of the susceptibility χ and of the product χT with changing temperature!
4. What is the difference between the super-exchange mechanism and spin polarization? What are the similarities?
5. Why is there no octahedral iron(II) complex with a ligand field splitting of 15,000 cm^{-1}?
6. Why is the change in bond lengths during a spin crossover most pronounced for iron (II) compounds and weakest for cobalt(II) compounds?
7. Why are the nickel(II) complexes [Ni(L)$_2$] with L $=$ ketoenolate, a bidentate chelate ligand diamagnetic as monomers and paramagnetic as trimers?

Luminescence of Metal Complexes

11

In the chapter on the color of complexes, we looked at how a complex is excited from the electronic ground state to an electronically excited state. The different options of electronic transitions were discussed and the associated selection rules were discussed. We did not consider what happens to the high-energy excited state. There are three basic possibilities here: (i) it can return back to the ground state without radiation (recombination); (ii) it can release the energy again by emitting radiation (luminescence); or (iii) it can be used for reactions (See Chap. 13/Photocatalysis). The basics for the first two options will be discussed in the following chapter. The following review articles and books are recommended as further reading on this very current field of research [75–79].

11.1 Basics

Luminescent materials and the excited states involved are highly interesting and have a high application potential, especially with regard to the renewable energy politics. Long-lived excited states are important for the function of dye-sensitized solar cells (DSSC) in the conversion of light energy into electrical energy. Conversely, luminescent materials are needed for the energy-efficient conversion of electrical energy (current) into light in organic light-emitting diodes (OLED) or phosphorescent light-emitting diodes (PhoLED). Other areas of application are, for example, in imaging or as sensor materials.

The generation of the excited state essential for luminescence can occur not only by light absorption (also called photoluminescence), but also by an electric field (in LEDs) or a chemical reaction (chemiluminescence, e.g. luminol for the detection of blood). In the latter case, chemical reactions in living organisms are separately referred to as bioluminescence (e.g., fireflies). In the following, we will only consider photoluminescence and start with

© The Author(s), under exclusive license to Springer-Verlag GmbH, DE, part of
Springer Nature 2023
B. Weber, *Coordination Chemistry*,
https://doi.org/10.1007/978-3-662-66441-4_11

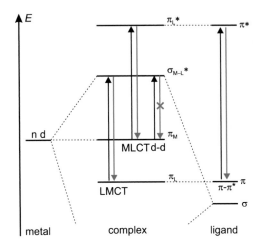

the various electronic transitions in complexes, which we have already learned about for absorption, and which can take place in the same way vice versa for emission.

In Fig. 11.1, a simplified general MO scheme of an octahedral complex is given, as discussed already in the chapter on bond theories. In general, four different types of transitions can occur in metal complexes, which, as noted in the chapter on color of complexes, differ characteristically in their probability and energy. The d-d transitions, also called metal-centered (MC) transitions in the following, are symmetry forbidden (parity forbidden) in inversion-symmetric complexes according to Laporte. By oscillations with u-character, the inversion center can be temporarily removed and the transitions can still take place, but with low intensity. In this context, one also speaks of vibronic coupling, since electronic and vibrational transitions are interrelated and influence each other. In the presence of suitable ligands, ligand-centered (LC) transitions between occupied and empty ligand orbitals are also possible (e.g. $\pi - \pi^*$). In addition, there are metal-to-ligand and ligand-to-metal charge-transfer transitions (MLCT and LMCT), and in some cases ligand-to-ligand charge-transfer (LL'CT) transitions. Theoretically, the reverse process, spontaneous return to the ground state under emission of light, is now conceivable for all transitions. This prediction does not hold for several reasons. Rather, luminescence (if it occurs; the vast majority of complexes do not emit) generally does not occur from the initially excited state, but from the lowest excited state. This means that before emission, relaxation processes of a structural and/or electronic nature occur. Only the lowest excited state is long-lived enough to allow radiative and non-radiative processes to compete. This fact is well known as *Rule of Kasha* from organic photochemistry. It should be noted that this rule allows more exceptions in complex chemistry than in organic chemistry.

Consequently, this statement means that the nature of emission is largely decoupled from the nature of excitation. While we may be able to selectively occupy all of the states described above in the excitation, it is ultimately only the lowest excited state that controls the emission. A further aggravation applies when the lowest excited state is d-d in

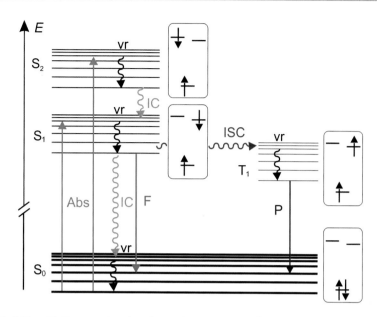

Fig. 11.2 Jablonski diagram showing the various processes that can take place after electronic excitation of a molecule [76, 78]. Abs: absorption; F: fluorescence; P: phosphorescence; vr: vibrational relaxation; IC: internal conversion; ISC: intersystem crossing. For a better illustration of the different electronic states, corresponding microstates are given in the boxes

character. No emission is usually observed for the metal-centered d-d transitions. We consider an exception to this rule at the end of the chapter for the chromium(III) ion.

When considering the luminescence of complexes, we now investigate what is the fate of the excited state. What is its evolution with time, i.e., what are the ways and means of returning to the electronic ground state? Figure 11.2 summarizes the various processes [76, 78]. In the chapter on the color of complexes, we have always assumed that the return to the electronic ground state is radiationless, i.e. the energy of the excited state is released by vibrations (vibrational relaxation, vr) or collision with other molecules. Luminescence is the general term for processes in which the energy of the electronically excited state is released by emission of radiation. A distinction is made between the spin-allowed process of fluorescence and the spin-reversed, and thus formally spin-forbidden, phosphorescence. Fluorescence is often observed in organic molecules, involving the transition between an S_1 (or S_n) excited state to the S_0 ground state. S stands for singlet and refers to the spin multiplicity of the system. The non-radiative relaxation between two excited states under spin conservation is called internal conversion (IC). For metal complexes with heavy transition metals and pronounced spin-orbit coupling (see magnetism), an additional transition known as intersystem crossing (ISC), e.g., from a singlet to a triplet state, can occur. As indicated in Fig. 11.2, triplet states are always energetically below the corresponding singlet states for the same electron configuration due to the larger exchange interaction. From here, spin-forbidden emission of radiation from an excited triplet state to the singlet ground state is possible, phosphorescence. Since this process is spin forbidden, the excited state lifetimes for phosphorescence are much longer than for fluorescence. Phosphorescence, unlike fluorescence, is quenched by oxygen, which also has a triplet ground state.

Decisive for which transitions take place are always the rates of the processes or the lifetimes of the individual states. For this purpose, we next consider the Jablonski diagram shown in Fig. 11.2, in which all the processes described are given. The associated lifetimes are summarized in the box Kinetics. During excitation, starting from an electronic and vibrational ground state (S_0), an electron is raised to an excited state with the same spin (S_1 or S_2). For this transition we must now take the Franck-Condon principle into account [80], which we already got to know for redox reactions but have so far disregarded in the case of the color of complexes. Since an electron is much lighter than an atomic nucleus, electronic transitions and nuclear motion are thought of to be decoupled. This means that the positions of the atomic nuclei of our complex (or molecule) do not change during the electronic transition; this is why we speak of a vertical transition, because we do not move on the reaction coordinate. However, excitation occurs not only to the vibrational ground state of the electronically excited state, but also to excited vibrational states. In the excited state, there is usually a different electron density distribution. For example, antibonding orbitals could be occupied with respect to certain bonds, leading to bond elongation at equilibrium. Therefore, after the very fast excitation (10^{-15} s), a vibrational relaxation (vr) takes place, in which the complex or molecule assumes the equilibrium geometry of the respective excited state. This process is also called vibrational cooling (vc).

If several excited states exist, they are often energetically close to each other and internal conversion (IC, no spin reversal) or intersystem crossing (ISC, spin reversal) can take place. Both processes are isoenergetic (hopping) processes between the potential energy surfaces with subsequent vibrational relaxation. In the following, we will take a closer look at the details of this using selected examples.

To compare the ratio of the individual processes, the lifetime τ' or the half-life τ of the excited state is often given. Both values refer to a (Boltzman distributed) collection of molecules and the processes considered here are monomolecular processes. While the half-life is defined as the time at which half of all molecules in our case, for example, have returned to the electronic ground state, the lifetime refers to the time where 1/e-th of the molecules are still in the electronic or vibrational excited state.

Typical lifetime (τ') of the individual processes [76]:

Absorption   Abs 10^{-15} s

Fluorescence   F 10^{-10} s – 10^{-7} s

Phosphorescence P 10^{-6} s – 10 s

Vibrational relaxation   vr 10^{-14} s – 10^{-10} s

Intersystem crossing   ISC 10^{-12} s – 10^{-8} s

11.2   Fluorescence Using the Example of a Zinc(II) Complex

Figure 11.3 on the right shows the absorption and emission spectrum of a square pyramidal zinc(II) complex. It is noticeable that absorption and emission behave like image and mirror image. This is due to the fact that vibrational excitations play a role in the electronic transition in both cases. We take this circumstance into account in an extended Jablonski diagram (Fig. 11.3), which uses the Zn harmonic oscillator to describe the diagnostic oscillation modes involved in excitation and de-excitation. Furthermore, it is noticeable that the absorption and emission spectra are shifted against each other on the energy scale. Fundamentally, the emitted quanta have a lower energy (longer wavelength) than the absorbed ones. The energy difference is called the Stokes shift. The origin and magnitude of the Stokes shift can be easily interpreted in the context of the vibronic coupling of the electron transition (see below). Usually, the electron distributions in the electronic ground state and in the electronically excited state differ, and so do the potential hypersurfaces and

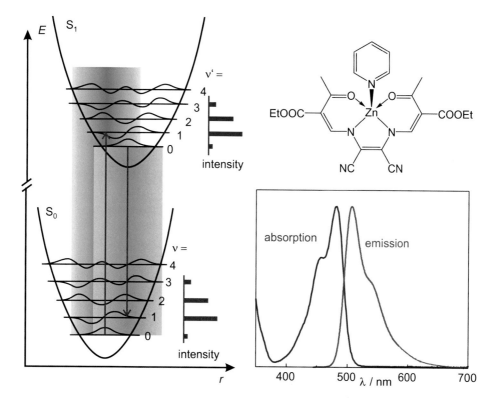

Fig. 11.3 Left: extended Jablonski diagram for a fluorescent molecule, for the representation of the vibrational levels and their eigenfunctions the harmonic oscillator was used as an approximation. Right: Structure of the zinc complex [Zn(L1)py] with normalized absorption and emission spectrum. The energy spacings between the individual bands of absorption and emission of approx. 1350 cm^{-1} correspond to framework deformation vibrations of the ligand

the location of their absolute minima. The excitation causes the occupation of (partially) antibonding states. This leads to the loosening of one or more bonds. This loosening is accounted for by the fact that the minima of the excited state and the ground state are shifted with respect to each other on the reaction coordinate r; the normal case of a bond loosening in the excited state then means a shift to the right, r(GS) < r(ES). Electronic excitation has many parallels to the Markus theory for electron transfer processes; it is also an electronic transition. As a first approximation, we assume that the transitions are ligand-centered. The zinc(II) ion plays a minor role in the electronic transitions as a formally non-redox active diamagnetic d^{10} element. To understand the different position and shape of the bands for absorption and the corresponding emission, we need the energy diagram given on the left. Each electronic state of a complex or molecule has a potential hypersurface whose absolute minimum corresponds to the equilibrium geometry in the molecule for this state. The total energy of the electronic state (here the S_0- or S_1-state) is a function of the electron distribution (i.e. the electron configuration or more precisely the individual microstates) and the nuclear distances and is determined by the core-electron, core-nucleus and electron-electron interactions. The potential hypersurface describes the energy change of the state as a function of any change of geometry. For a better representation, one does not consider the whole 3 N-dimensional space, but selects a reasonable normal coordinate with the associated vibrational states. These can be approximated by a harmonic or the Morse potential (anharmonic). In Fig. 11.3 on the left, a harmonic potential is given (harmonic oscillator, equal distances between the individual oscillation levels, equilibrium distance does not change), while in Fig. 11.2 the Morse potential was assumed (distances between the individual oscillation levels decrease with increasing energy). Since the difference between the two models is negligible for the first oscillation levels considered in the following, we use the simpler harmonic oscillator model. We assume that for the electronic ground state only the vibrational ground state ($\nu = 0$) is relevant for the excitation (at room temperature this is a reasonable assumption), while in the excited state several vibrational levels have to be considered. The intensity of the individual vibronic transitions (S_0, $\nu_0 \rightarrow S_1$, ν_n' with $n = 0, 1, 2, 3, \ldots$; shorter: $S(00) \rightarrow S(1n)$) is proportional to the square of the overlap integral between the vibrational eigenfunctions of the states involved in the transition. If the equilibrium distance of the atomic nuclei does not change during the electronic transition (both potential wells lie exactly on top of each other), then the transition between the two vibrational ground states ($S(00) \rightarrow S(10)$) is most intense. When the equilibrium distances change significantly, as shown in Fig. 11.3, the most intense transitions with the largest overlap integral are those leading to higher vibrational levels ($S(00) \rightarrow S(1n)$). This leads to an asymmetric broadening of the absorption band, where the vibrational fine structure is often resolved. This is the case in the example shown here. Note that excitation can occur from all possible core distances of the vibrational ground state, the range is highlighted in blue in Fig. 11.3.

11.2.1 Prerequisites for Fluorescence

The molecule in the electronically excited state can release its energy again (i.e. relax) via various possibilities. The first step is usually vibrational relaxation. It includes intramolecular radiationless processes, i.e. energy dissipation to different acceptor modes in the molecule. If the energy is also transferred to surrounding molecules, it is called intermolecular vibrational energy transfer, and the environment of the excited molecule is heated. To consider intramolecular processes, we must first note that a nonlinear molecule (or complex) of N atoms has $3N - 6$ degrees of vibrational freedom. In a three-atomic linear molecule like CO_2 3N - 5 vibrations are allowed, so three so-called normal vibrations are to be expected. In larger molecules, the number of allowed vibrations increases correspondingly rapidly. For the example of the Zn complex in Fig. 11.3 with 58 atoms, for example, there are already 168 normal vibrations. However, not all of these can be excited during an electronic transition. To understand this, let us consider again d-d transitions in an ideal octahedral complex. In the chapter on the color of coordination compounds, we learned that d-d transitions are forbidden in centrosymmetric complexes according to Laporte. The fact that color is observed nevertheless is due to the fact that centrosymmetry can be cancelled by vibrations. This condition is fulfilled for normal oscillations with odd parity of the oscillation mode. There is then a simultaneous excitation of the electronic transition and the odd normal vibrational mode, this is called vibronic coupling. In such coupled transitions, the shape of the bands of the electron spectrum depends on the nuclear motion in the excited electronic state (absorption) and in the ground electronic state (emission). Thus, if the ground state and the excited state are structurally very different, an additional vibrational progression takes place. Here oscillations of not symmetry-allowed vibrational levels are also excited. The excess vibrational energy is distributed during relaxation to other, isoenergetic vibrational states that were not excited. The time to relax to the vibrational ground state is about $10^{-10} - 10^{-14}$ s. These steps always take place after electronic excitation. In the equilibrated excited vibrational state (S (10)), mechanistic bifurcation now occurs into competing processes, nonradiative deactivation (nr), radiative deactivation (fluorescence), and nonradiative spin transition (intersystem crossing). The efficiency of the fluorescence $\phi(F)$ is measured by the respective rate constants k:

$$\phi(F) = \frac{k(F)}{k(F) + k(ISC) + k(nr)}$$

The emission spectrum observed after excitation and relaxation is always shifted towards longer wavelengths (lower energies), as shown in Fig. 11.3 on the right. Fluorescence again involves excitation of several vibrational levels, this time of the ground state, with the intensity again depending on the square of the overlap integral of the vibrational eigenfunctions involved. The rules for emission are the same as for absorption. Fluorescence is always preferred when nonradiative channels become inefficient, i.e., in the absence of heavy atoms ($k(F) > > k(ISC)$) and when the energetic separation of S_0 and S_1 is large. The last point, $k(nr)$ is inversely proportional to

$$E(S_1) - E(S_0)$$

, is also called energy-gap-law. In addition to a small energy difference between the electronic ground state and the electronically excited state, a strong distortion of the electronically excited state can also be a reason for efficient radiationless relaxation (strong-coupling limit).

Experience shows that it are mostly planar, less flexible molecules that fluoresce efficiently. Why this? This finding can again be satisfactorily interpreted in the parabolic picture of vibronic coupling shown in Fig. 11.3. In the case of rigid molecules, excitation causes only a relatively small structural adjustment to the new electronic state; the two parabolas are only slightly shifted with respect to each other. Accordingly, there is no coupling from the vibrational ground state of S_1 to excited vibrational levels of S_0; the curves intersect only at highly excited vibrational states of S_1 (hot states). The situation is different when the potential wells are strongly horizontally displaced. The curves of S_0 and S_1 intersect close to the minimum of S_1; an activation-free radiationless depopulation of the excited state occurs.

11.3 Phosphorescence of Diamagnetic Complexes

11.3.1 d^6 [M(bipy)$_3$]$^{2+}$ Complexes

As a first example for complexes with long-lived MLCT states responsible for phosphorescence, we consider the [Ru(bipy)$_3$]$^{2+}$ ion and compare the situation with the lighter homologue [Fe(bipy)$_3$]$^{2+}$, which shows neither phosphorescence nor fluorescence. Low-spin d^6 complexes such as ruthenium(II) complexes with various (poly)pyridine ligands and iridium(III) complexes have long-lived excited states with CT character. The combination of these two factors (long-lived + CT character) is the prerequisite for the use of such complexes in dye-sensitized solar cells, PhoLEDs, and photocatalysis [77–79]. The longevity stems from the fact that the lowest excited state is a triplet state whose deactivation requires spin reversal.

Phosphorescence, as stated above, is a radiative transition between states of different spin multiplicity. Since the formation of the phosphorescent state also requires a spin reversal, heavy atoms, for example, promote both population and depopulation of the emitting state via the spin-orbit interaction. Quite analogously to fluorescence, there is also a dependence on the energy separation here

$$E(T_1) - E(S_0)$$

of the states involved in terms of the energy gap law.

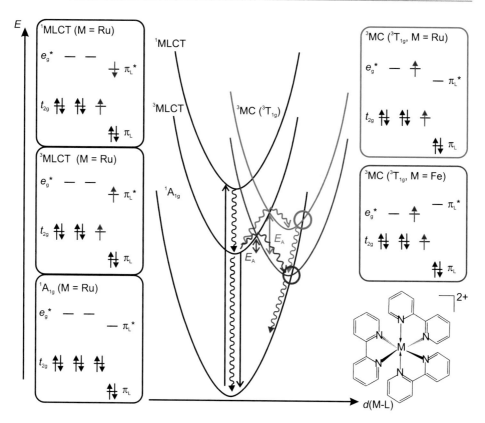

Fig. 11.4 Schematic representation of the potential hypersurfaces of a $[M(bipy)_3]^{2+}$ complex with M = Ru or Fe and selected microstates. Excitation of the complexes occurs for both metals from the $^1A_{1g}$ ground state to a ^1MLCT state, which transitions to a ^3MLCT state by fast ISC. Phosphorescence occurs from this state. In competition, thermally activated IC causes a transition to the ^3MC state, which has a crossing point with the $^1A_{1g}$ energy potential surface over which rapid thermal relaxation occurs. The relative position of the two potential hypersurfaces (^3MLCT and ^3MC) to each other depends on the energetic order of the e_g^* and π^* orbitals and is crucial for whether phosphorescence is observed or not

To understand the fundamentals of phosphorescence at $[Ru(bipy)_3]^{2+}$, in addition to the location of the potential hypersurfaces, we consider the relevant microstates, i.e., individual electron configurations that help us better illustrate the basic principle of phosphorescence. In Fig. 11.4, the potential hypersurfaces and the electron configurations of individual microstates relevant for the following discussion are given. We will not show the vibrational levels for the potential hypersurfaces in the following examples.

$[Ru(bipy)_3]^{2+}$ is a diamagnetic low-spin complex with a $^1A_{1g}$ ground state. Excitation with light causes an MLCT to a ^1MLCT state (by this we mean the excited state, the unpaired electron is in the π^* orbital of the ligand and an electron hole on the metal, formally [Ru^{III}(bipy$^{\cdot-}$(bipy)$_2$]), followed by very fast ISC to the ^{3}MLCT state. The excitation hardly changes the Ru-N lengths (see also reorganization energy during electron transfer between [Ru(bipy)$_3$]$^{2+}$ and [Ru(bipy)$_3$]$^{3+}$ in the chapter Redox reactions). Accordingly, the three potential hypersurfaces are hardly shifted with respect to each other along the x-axis, that corresponds to the M-L distance. The negative charge of the bipyridine radical anion formally formed in the excited states is initially delocalized across the three ligands, but subsequently localizes to one of the bipy ligands. For the [Ru(bipy)$_3$]$^{2+}$*, the thermally equilibrated ^{3}MLCT state is the lowest excited state. It has a microsecond lifetime at RT. Next to the ligand-centered triplet state, there is also a metal-centered triplet state (^{3}MC) where the higher-energy unpaired electron is not localized in the π* orbital of the ligand, but in the e$_g$* orbital of the metal. In this case, the occupation of the σ-antibonding orbitals leads to a significant lengthening of the metal-ligand distance (cf. redox reactions, MO scheme of complexes). As a consequence, the potential hypersurface is shifted towards longer M-L distances, as shown in Fig. 11.4. Thermally activated IC can now take place between the metal- and ligand-centered triplet states. The efficiency of this process depends on the relative position of these two potential hypersurfaces with respect to each other, and hence on the energetic position of the e$_g$* and π* orbitals. For the 4d element ruthenium, the large splitting of the d orbitals in the octahedral ligand field results in the π* orbitals being energetically below the e$_g$* orbitals. Therefore, the ^{3}MLCT state is energetically more favorable than the ^{3}MC state and a relatively high activation energy is required for IC. In the case of the 3d element iron, the order of the orbitals is reversed and thus the ^{3}MC state is energetically lower than the ^{3}MLCT state. This leads to a much lower activation energy for the IC and a fast and efficient occupation of the ^{3}MC state. The change in distance and accompanying shift of the ^{3}MC energy potential surface along the x-axis creates a crossing point with the potential surface of the ^{1}A$_{1g}$ ground state. This crossing point, which opens another path for a radiationless relaxation, is called the minimum energy crossing point (MECP). It is very well seen in the schematic diagram that the location of the ^{3}MC (^{3}T$_{1g}$) potential has a great influence on the activation energy for the IC and the MECP point with the potential hypersurface of the ground state. This theoretical consideration agrees well with the experimental facts, that is, the emission quantum yield (ratio between the number of emitted and absorbed photons). For [Ru(bipy)$_3$]$^{2+}$, the emission quantum yield in solution at room temperature is 0.095 (without oxygen) and 0.018 (with oxygen, phosphorescence is quenched proportionally), respectively. Since the thermally activated IC contributes significantly to the nonradiative relaxation, the quantum yield and hence the lifetime of the excited triplet state is temperature dependent. The lifetime is 5 μs at 77 K and 850 ns at RT. For the homologous iron(II) complex [Fe(bipy)$_3$]$^{2+}$, the IC is so fast that no phosphorescence is observed even at low temperatures. Iridium(III) as a 5d center in a high oxidation state has a very large octahedral splitting Δ_O and thus a very high activation energy for the IC. Here this relaxation path is clearly disadvantaged and thus the emission quantum yield is once again higher.

11.3.2 Copper(I) Complexes as Example of 3d^{10}

One of the reasons for the interest in 3d element photochemistry is their generally much larger and wider availability. For example, the 3d element copper is much more readily available and less expensive than 4d and 5d elements such as ruthenium and iridium [75, 81]. When an electron-rich metal center such as the d^{10} copper(I) ion is combined with electron-poor ligands such as pyridines, MLCT states are occupied by excitation with light. Just as for the ruthenium complex discussed earlier, the metal is formally oxidized and the corresponding reduced radical anion ligand $L^{\cdot-}$ (e.g. py$^{\cdot-}$ or phen$^{\cdot-}$) is formed. An advantage over the ruthenium system is that no low-lying metal-centered (MC) states are present for d^{10} elements. Thus, tetrahedral copper(I) complexes with (poly)pyridines as ligands and MLCT states of suitable energy are ideal for the observation of long-lived excited states that may exhibit luminescence, as we have already encountered with the d^{10} ion zinc. The Zn/Cu difference can be determined by the character of the lowest excited state. In the case of Zn, only small metal contributions are observed, while in the case of Cu, the MLCT character dominates. Even though no direct relaxation via MC states can take place in the case of Cu(I), an efficient relaxation pathway exists. Massive structural reorganization reflects the Cu(II) character of the excited state and the preference of Cu (II) for the planar environment. This structural relaxation is called *excited state flattening distortion*. This distortion unfolds its effect via the horizontal shift of the potential curve of the excited state. The reason for this is the very different preferred coordination geometries of copper(I) and (II) at coordination number 4, which have already been discussed in the section on redox reactions. While copper(I) has no configurationally preferred geometry and thus coordinates tetrahedrally at coordination number 4, in the electronically excited state there is formally a copper(II) ion which, as a d^{9} ion, prefers Jahn-Teller geometries such as a square planar coordination environment. For complexes with monodentate or bidentate ligands that are free to move, flattening on the picosecond time scale occurs in the excited state, even in the ^{1}MLCT state, before the transition to the ^{3}MLCT state occurs. As shown in Fig. 11.5 on the left, highly distorted excited states (^{1}MLCT*) favor nonradiative relaxation by a significant shift of the potential hypersurfaces along the reaction coordinate, which this time represents the angle α between the planes of bidentate ligands, 90 + x. To obtain efficient luminescence, so-called *nested states* are required in the excited state, which are only slightly distorted with respect to the ground state. In solution, this requirement can be realized if chelating ligands with sterically demanding substituents are used, which prevent the flattening distortion in the excited state. An alternative is to characterize the complexes as solids, since here the structure is fixed by the packing.

Complexes of the composition [CuX(PPh$_3$)$_2$(L)] (L = nitrogen donor such as pyridine, X = halide) show intense luminescence in the solid with a quantum yield of up to 100%; in solution these complexes do not emit. Figure 11.5, right, shows the Jablonski diagram and microstates for the complex [CuI(PPh$_3$)$_2$(py)]. For this and similar copper(I) complexes, different emission properties are observed as a function of temperature. As already stated, the excitation takes place analogously to the low-spin d^{6} systems discussed earlier into

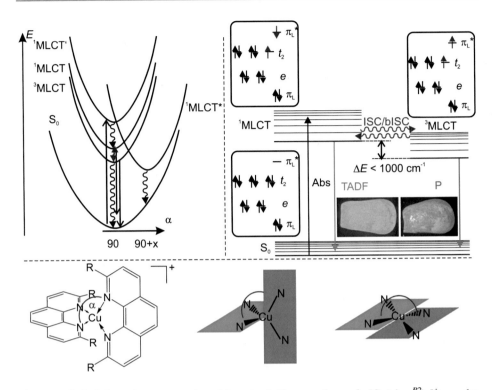

Fig. 11.5 Left: Schematic representation of the potential hypersurfaces of a $[Cu(phen^{R2})_2]^+$ complex in the ground and excited states. Upon excitation, a copper(II) complex is formally formed, leading to planarization of the excited-state structure (increase in the dihedral angle α between the two phenanthroline ligands) and consequent nonradiative deactivation. The corresponding structures with angle α are given below. Right: Jablonski diagram of complex $[CuI(PPh_3)_2(py)]$ with corresponding microstates. Fluorescence is observed at room temperature, while phosphorescence dominates at low temperatures. The different color of emission as a function of temperature is shown

excited ^{1}MLCT' states. Via vibrational relaxation and IC, the lowest excited ^{1}MLCT state is reached, from which ISC into the lowest-energy ^{3}MLCT excited state takes place. Thus, all the conditions for the occurrence of phosphorescence are fulfilled, which can also be observed at low temperatures. As a spin forbidden transition, phosphorescence is comparatively slow and the lifetime of the ^{3}MLCT state is correspondingly long. For the copper complex $[CuI(PPh_3)_2(py)]$ and similar derivatives, there is another special feature. The ^{3}MLCT state is energetically very slightly below the ^{1}MLCT state, (<1000 cm^{-1}, <12 kJ mol^{-1}). The combination of the small energy gap and the long lifetime of the ^{3}MLCT state enables thermally activated back-ISC (back-intersystem crossing) to the ^{1}MLCT state at sufficiently high temperatures (e.g. RT). From here, spin-allowed fluorescence can now take place, which is much faster and thus dominates at higher temperatures. Fluorescence is thus observed with a decay time equal to the lifetime of the triplet state. The whole process is called *thermally activated delayed fluorescense* (TADF), in which the occupation of ^{1}MLCT and ^{3}MLCT is temperature dependent and follows a Boltzmann distribution.

11.4 Phosphorescence of Metal-Centered Transitions

In the previous examples, CT states (mainly MLCT in our case, but there are also LMCT examples with d^0 and d^5 metals) were the cause for the observation of luminescence, while for metal-centered transitions we had previously assumed that non-radiative relaxation dominates. The reason for the absence of luminescent d-d states is that MC excited states lead to a significant geometry distortion in most cases. Normally, in octahedral complexes, the occupation of the antibonding $e_g{}^*$ orbitals (d-d transition) leads to a strong distortion in the excited state. In contrast, states with low geometric distortion *(nested states)* can lead to the emission of light. It is, thus, obvious that luminescent MC states will be found where the occupation of the $e_g{}^*$ orbitals does not change despite excitation. Excited states without a change in the occupation of the t_{2g} and $e_g{}^*$ orbitals can be obtained if one of the electrons reverses its spin. The so-called *spin-flip* states obtained in this way are usually long-lived *nested states* from which phosphorescence can be observed [75, 82].

This condition is achieved with a d^3 electron configuration, such as for the chromium (III) ion in an octahedral ligand field. Figure 11.6 shows the simplified Tanabe-Sugano diagram associated with a d^3 system and a Jablonski diagram showing the states relevant to the discussion. Starting from the quartet $^4A_{2g}$ ground state, there are quartet ($^4T_{2g}$) and doublet (^2E_g and $^2T_{1g}$) excited states in a similar energy range. The doublet states are *spin-flip* states for which the occupation of the t_{2g} and $e_g{}^*$ orbitals does not change, but only the spin of an electron reverses. The energy of these undistorted or only very weakly distorted states is therefore almost independent of the ligand field splitting Δ_O. In contrast, the energy of the excited quartet states clearly depends on the ligand field splitting Δ_O. This leads to the fact that if the ligand field splitting is sufficiently large ($\Delta_O >> B$), ISC can occur from the excited quartet state to the doublet state (excited state with lowest energy). If the energetic difference is large enough that no bISC can occur (unlike the copper complexes just discussed), then a sharp phosphorescence band is observed. We start with an example from solid-state chemistry where this principle is manifested. In ruby (Al_2O_3 doped with Cr(III)), two sharp phosphorescence bands are observed at 694.3 nm and 692.9 nm, the so-called R_1 and R_2 lines. The chromium(III) ion here has an almost ideal octahedral coordination environment and the chromium-oxygen distances in the corundum host lattice are much shorter than would be the case for free Cr(III) complexes with O_6 environments. The compression of the coordination sphere is a direct consequence of the steric constraints of the host lattice. In other words: What does not fit (r(Cr) > r(Al)) is made to fit. Thus, the rules for the realization of molecular analogues are already set. A strong ligand field can be realized by cyanido ligands, for example. In addition, an octahedral geometry as ideal as possible is needed, which on the one hand leads to a larger splitting Δ_O, and in addition suppresses the non-radiative relaxation, which is supported by

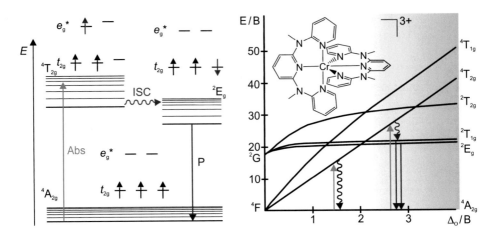

Fig. 11.6 Left: simplified Jablonski diagram of an octahedral chromium(III) complex with associated microstates. Right: simplified Tanabe-Sugano diagram for d^3 [ML$_6$] complexes. To observe phosphorescence, the ^2E_g and $^2T_{1g}$ states must be energetically below the excited state $^4T_{2g}$. Since the location of these two states is almost independent of the octahedral splitting Δ_O, this requirement can be realized if the ligand field splitting is sufficiently high. The suitable region is highlighted in color

distortion. Tridentate pyridine-based ligands can also lead to good emission properties if ligand design ensures an ideal octahedral geometry (L-M-L angle of 90°). This can be realized by spanning chelate-6 rings instead of chelate-5 rings.

In chromium(III) complexes, two doublet states are close together, as shown in Fig. 11.6. The energetic separation between the two states is relatively small. If the IC between the two states is fast and the two states are quasi-degenerate, then the states appear phenomenologically as one state with a common decay time. Under this condition, there is a Boltzmann-distributed occupation of the two states, and accordingly two emission bands are observed whose relative intensity depends on the temperature (the lifetimes of the two states are the same). This observation bears some resemblance to the temperature-dependent switch between phosphorescence and fluorescence discussed for copper (I) complexes, where the bISC can only take place if the ^{3}MLCT state lives long enough. In both examples, the prerequisite is a weak distortion of the excited states. An application of such molecular chromium(III) complexes as optical thermometers e.g. in biological systems, which do not need an external reference, is conceivable.

11.5   Luminescence by Aggregation of Platinum(II) Complexes

In the examples we have discussed so far, the luminescent properties were based on the photophysical properties of a single complex or molecule. Finally, we consider an example where the luminescence is produced or significantly affected by the aggregation of single

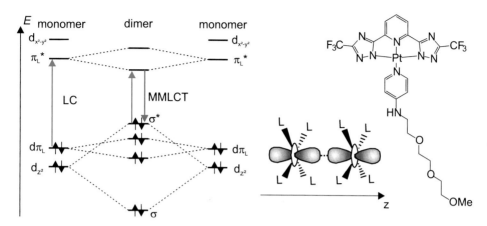

Fig. 11.7 Influence of metallophilic interactions in square planar platinum(II) complexes on the relative position of the orbitals and thus the photophysical properties. As an example, the structure of a neutral square planar platinum complex is given, which shows a very diverse aggregation behavior due to its structure [85]

molecules into a supramolecular aggregate. For this we consider square planar platinum (II) complexes [83, 84]. Figure 11.7 shows the location of the relevant orbitals for a mononuclear and a dimeric complex. The energy schemes shown require square planar platinum(II) complexes with multidentate strong-field ligands, which are good π-acceptors. The strong-field ligands cause an energetic rise of the $d_{x^2-y^2}$ -orbital and thus of the metal-centered states, while the π-acceptor ligands provide low-lying emissive MLCT states. In the mononuclear compounds, the highest occupied molecular orbital (HOMO) has $d\pi$-character and the lowest unoccupied molecular orbital (LUMO) has π^* character. Both orbitals are primarily ligand centered, and the electronic transitions occurring between the two orbitals have LC character accordingly. The d_{z^2} orbital is fully occupied in square planar d^8 systems and has little interaction with ligands. With the given framework, it is below the HOMO for the mononuclear complex, but is certainly available for MLCT transitions. The order of the orbitals (metal-based vs. ligand-based) changes upon interactions with additional ligands coordinating in axial position at the platinum or by intermolecular interactions with neighboring platinum complexes. In particular, the occurrence of platinum-platinum interactions (metallophilic interactions) leads to a new HOMO with σ^* character corresponding to the antibonding orbital of the metallophilic interactions between the two d_{z^2} orbitals, as shown in Fig. 11.7. The reason for the aggregation of the monomers is not the metallophilic interaction (since binding and antibinding orbitals are fully occupied, the bond order is 0), but π -π interactions between the planar ligands and, in addition, dispersive interactions of the substituents at the ligand. For example, alkyl chains can affect aggregation via van der Waals interactions. The new energetic order of the orbitals then leads to the occurrence of metal-to-metal-to-ligand charge transfer transitions (MMLCT, $d\sigma^*-\pi^*$). The occupation of the underlying CT states requires less energy than for the monomers, and is thus bathochromically shifted in the spectrum (redshift to the longer wavelength lower energy region of the spectrum). The luminescence from these states is correspondingly bathochromically shifted relative to that of the monomer. As can be seen in Fig. 11.7, the energetic position of the $d\sigma^*$ orbital, and hence the energy of the MMLCT, depends strongly on the Pt-Pt spacing and hence the strength of the dispersive interactions. Typically, the interactions occur at distances shorter than 3.5 Å. The aggregation and thus the luminescence can be influenced by the environment, e.g. the solvent. A nice example is the complex shown in Fig. 11.7 from the publication [85], where different aggregates with different emission properties are formed. In the Supporting Information, one can watch the videos on this from the publisher.

11.6 Questions

- With the help of a Jablonski diagram, explain the following terms: Fluorescence, phosphorescence, internal conversion (IC), intersystem crossing (ISC), thermally activated delayed fluorescence (TADF)!
- What is meant by vibrational relaxation? To what extent does it influence the quantum yield of the emission? In this context, explain the term *nested state* using a self-selected example.

Bioinorganic Chemistry

In biological systems, metals play an essential role. In proteins (= metalloproteins), the metal centers are either directly bound to the protein scaffold via amino acid side groups or coordinated by macrocyclic ligands. Depending on the function of the metal centers, metalloproteins can be divided into five basic types.

- *Structure formation:* Metal ions bound to a protein can co-determine and stabilize the tertiary or quaternary structure. Examples include zinc fingers (Zn^{2+}) or proteins stabilized by alkaline earth metal ions in thermophilic bacteria. These exist under conditions where protein denaturation would normally occur.
- *storage of metal ions:* An impressive example here is ferritin, the storage protein for iron in higher organisms. An inorganic iron(III) oxide core is surrounded by a protein shell.
- *Electron transfer:* Metal centers coordinated to protein scaffolds or macrocycles play an important role in the electron transfer chains of respiration and photosynthesis.
- *Binding of oxygen:* Here the metal centers are either directly bound to the protein scaffold or localized in a porphyrin unit.
- *Catalysis:* This group of metalloproteins is called metalloenzymes and can be further subdivided according to the type of reaction.

The systems found in nature have been optimized by high evolutionary pressure. Understanding the functioning of metalloproteins and their peculiarities helps us to learn from nature's model, i.e. to practice biomimetic inorganic chemistry. Within the scope of this book, two selected examples are discussed in detail. The peculiarities of coordination chemistry are highlighted and the importance of model compounds is discussed. The books "Bioinorganic Chemistry" by Kaim and Schwederski [86] and "Bioinorganic

Chemistry: Metalloproteins, Methods and Concepts" by Herres-Pawlis and Klüfers [87] are recommended for further reading.

12.1 Biologically Relevant Iron Complexes

Iron is an essential trace element for almost all organisms. Its distribution in the body of an adult human is given in Table 12.1. From it, the versatile role of iron and especially of the heme group in human biochemistry becomes obvious. Since oxygen transport is not a catalytic but a "stoichiometric" function, about 65% of the iron found in the human body is accounted for by the transport protein hemoglobin, and myoglobin accounts for about 6%. Metal storage proteins such as ferritin essentially make up the rest of the body's iron, and the catalytically active enzymes are naturally present only in small quantities.

Many, but not all, of the redoxcatalytical active iron enzymes contain the heme moiety in the same way as hemoglobin and myoglobin. The hemoproteins include peroxidases, cytochromes, cytochrome c oxidase and the P450 system. The list shows the determining role that the protein environment apparently plays in the different functionality of a tetrapyrrole complex. Heme-containing enzymes are involved in electron transport and accumulation, in the controlled conversion of oxygen-containing intermediates such as O_2^{2-}, NO_2^- or SO_3^{2-}, and, together with other prosthetic groups, in more complex redox processes. One prosthetic group belongs to the cofactors. A cofactor is a molecule or group of molecules that is essential for the function of a particular enzyme. A distinction is made between a prosthetic group that is covalently bound to the enzyme and a coenzyme that is not covalently bound and can dissociate again after the reaction.

Table 12.1 Selected biologically relevant iron complexes and their function in the human body. The quantities given refer to the distribution in the body of an adult human being

Protein	Amount of iron (g)	% of total iron	Heme (h) or non-heme (nh)	Function
Hemoglobin	2.60	65	h	O_2-transport in the blood
Myoglobin	0.13	6	h	O_2-storage in the muscle
Ferritin	0.52	13	nh	Cellular iron storage
Hemosiderin	0.48	12	nh	Cellular iron storage
Catalase	0.004	0.1	h	Metabolism of H_2O_2
Cytochrome c	0.004	0.1	h	Electron transfer
Cytochrome c oxidase	<0.02	<0.5	h	Terminal oxidation
Flavoprotein oxygenases (P450)	Low	Low	h	Incorporation of molecular oxygen
Iron-sulfur proteins	Approx. 0.04	Approx. 1	nh	Electron transfer

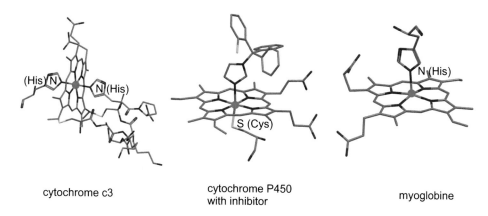

cytochrome c3

cytochrome P450
with inhibitor

myoglobine

Fig. 12.1 Heme unit in cytochrome *c3*, cytochrome P450 and myoglobin. The different functions (electron transfer, catalysis, oxygen storage) are determined by the additional ligands and the protein environment

The function assumed by the heme unit depends on the additional ligands at the iron center and the environment in the protein pocket. The heme unit consists of iron as the central ion surrounded by a tetradentate macrocyclic N_4^{2-} ligand, which is a conjugated aromatic system. This ligand, which occurs frequently in biological systems, is known as protoporphyrin IX. In Fig. 12.1, the heme units from cytochrome *c3*, cytochrome P450, and myoglobin are given as examples. In cytochrome *c*, which is responsible for electron transfer, both axial coordination sites of the iron are occupied by an imidazole ligand (histidine side chain from the protein). For rapid electron transfer (see Chap. 7), it is necessary that the coordination environment of the iron in the different oxidation states (+ 2 and + 3) does change as little as possible. This is well realized by the rigid macrocyclic ligand and the protein environment. The catalytically active cytochrome P450 and the myoglobin responsible for oxygen storage in muscle each have a free coordination site where oxygen can coordinate. The difference in function is determined by the protein environment and the sixth ligand (nitrogen from histidine in myoglobin or sulfide from cysteine in cytochrome P450).

12.1.1 Model Compounds

In hemoglobin and myoglobin, the iron center in the unloaded state is pentacoordinated with an imidazole (from the amino acid histidine) as the fifth ligand and is present in the +2 oxidation state. Upon reaction with oxygen, the system changes to a six-coordinated state. Whether the iron center is simultaneously oxidized to the trivalent state (with reduction of the coordinated oxygen to hyperoxide) is still controversial in some cases and will be considered in more detail below. Model compounds are often used to gain a better insight into the bonding properties. Figure 12.2 shows the heme unit and potential ligands for model compounds.

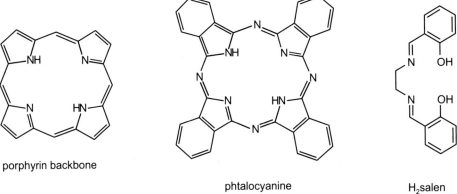

heme (Fe-protoporphyrin IX)

porphyrin backbone

phtalocyanine

H₂salen

Fig. 12.2 Schematic representation of the heme unit (top) and of various ligands for model compounds (bottom). The model compounds can structurally mimic the natural model (synthetic porphyrins, pthalocyanine ligands), but there are also model compounds that look quite different at first glance but reproduce the function of the natural model well

The advantage of model compounds is that they are better accessible and often easier to characterize. A distinction is made between structural and functional model compounds. As the name suggests, structural model compounds reflect the structure (and spectroscopic properties) of the metalloproteins, while functional model compounds reflect the function (e.g. catalytic property). An example of a structural model compound is given in Fig. 12.3. As for heme, the iron center is surrounded by a tetradentate macrocyclic N_4^{2-} ligand, but it is not fully conjugated. However, in this complex, reaction with oxygen (from the air) does not lead to reversible bonding, but to irreversible oxidation of the metal center to the trivalent oxidation state, forming a μ-oxido complex. The same behavior is observed for synthetic porphyrins (the heme basic structure with other substituents), so it can be

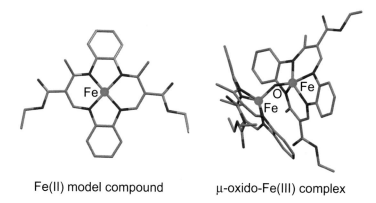

Fe(II) model compound   μ-oxido-Fe(III) complex

Fig. 12.3 Example of a structural iron(II) model compound for heme and its reaction product with oxygen

concluded that the surrounding protein is instrumental in the reversible character of the Fe-O$_2$ bond. Indeed, reversible binding of oxygen could be realized in so-called "picket fence" porphyrins, where the protein pocket is simulated by very bulky and spatially demanding substituents. The first functional model compound for this reaction was a cobalt salen complex.

12.2   Oxygen Transport Using the Example of Hemoglobin

Our Earth's primordial atmosphere consisted of less than 1% oxygen by volume. The emergence of photosynthesis led to a continuous increase in the oxygen concentration in the atmosphere. For the organisms living at that time, this was synonymous to an environmental catastrophe. Oxygen is the element with the second highest electronegativity (after fluorine) and a very strong oxidizing agent. The strong oxidizing character and the highly reactive (mostly radical) intermediates occurring in these reactions led to the fact that only organisms with suitable protective mechanisms survived the serious environmental changes. Further evolution led to novel organisms that used the reverse reaction of photosynthesis, controlled cold combustion, for energy production. This required the development of ways to absorb, transport, and store oxygen. As an example of the reversible binding of oxygen, the basic requirement for oxygen transport, hemoglobin is discussed in the following. This issue is not only interesting because of its biological necessity. Energy-intensive processes for the separation of oxygen from air, such as its fractional distillation, could be replaced by more gentle processes.

To better guide the following discussion, we first need to revisit the molecular properties of oxygen. Although oxygen is a very strong oxidant and the corresponding reactions are exothermic, a high activation energy is required for these reactions to proceed. The reason for this is the triplet ground state of the oxygen molecule, shown in Fig. 12.4 using the MO

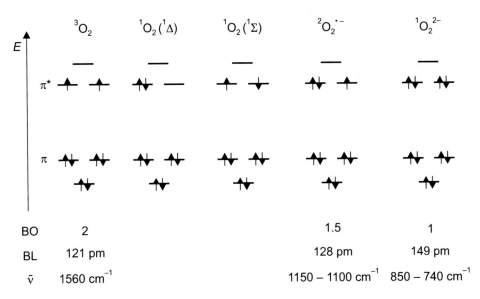

Fig. 12.4 MO scheme of oxygen ($^3O_2$ and $^1O_2$), the superoxide radical anion ($O_2^{\cdot-}$), and peroxide (O_2^{2-}). Only the orbitals relevant for the further discussion are shown. BO stands for bond order and BL for oxygen-oxygen bond length. In addition, the stretching vibration frequency is given

scheme. The two degenerate π^* orbitals are singly occupied according to Hund's rule. This leads to a paramagnetic ($S = 1$) ground state. The two excited singlet states are reached when the spin is reversed for one of the two electrons. Since these are excited states, Hund's rules lose their validity here! They are 90 kJ/mol ($^1\Delta$) and 150 kJ/mol ($^1\Sigma$) above the ground state. Because of the triplet ground state, reactions with other singlet molecules are inhibited (prohibition of spin reversal, reactions must proceed under spin conservation). The spin prohibition does not apply to reaction partners that themselves possess unpaired electrons. These include free radicals ($S = \frac{1}{2}$), excited triplet states ($S = 1$), and paramagnetic transition metal centers ($S \geq \frac{1}{2}$). The problem is that the reaction of most transition metals with oxygen is irreversible. Iron(II) ions present in aqueous solution are irreversibly oxidized to trivalent iron by atmospheric oxygen. These irreversible reactions are always accompanied by a cleavage of the oxygen-oxygen bond.

$$4\, Fe^{2+} + O_2 + 2\, H_2O + 8\, OH^- \rightarrow 4\, Fe(OH)_3$$

What Does $^1\Delta$ and $^1\Sigma$ Actually Mean for Singlet Oxygen?
As we have already learned about the term symbols for atoms and ions, there are also term symbols for the electron configuration of molecules. The number at the top left

(continued)

again stands for the multiplicity M of the term, which we also calculate for molecules from the total spin S according to the formula $M = 2S + 1$.

The capital Greek letter stands for the projection of the total angular momentum of the electrons onto the nuclear bond axis, Λ. For $\Lambda = 0, 1, 2, \ldots$ the letters Σ, Π, Δ, $\ldots$, are used. To determine Λ, we need the orbital angular momentum of the respective molecular orbital along the nuclear bond axis, λ, and then we need to sum up everything again. The orbital angular momentum of a σ orbital is 0, that of a π orbital is ± 1. For the determination of the term symbol, in analogy to atoms and ions, fully occupied orbitals are not considered. This means that in the case of oxygen, only the occupation of the two π^* orbitals that have an orbital angular momentum of $+1$ and -1 is of interest. If both π^* orbitals are occupied by one electron each, then.

$$\Lambda = +1 + (-1) = 0$$

and the term symbol is $^1\Sigma$. Incidentally, the same orbital angular momentum is obtained for triplet oxygen, which has the term symbol $^3\Sigma$. When both electrons are paired in the same π^* orbital, $\Lambda = 2$. Any state with $\Lambda > 0$ is doubly degenerate, here the values are $+2$ and -2, and the term symbol is $^1\Delta$. Incidentally, the orbital angular momentum of molecules also contributes to the magnetic moment of the compound (see the chapter on magnetism), and the $^1\Delta$ singlet oxygen is accordingly paramagnetic, although it has no unpaired electrons.

In the case of the superoxide radical anion, one π^* orbital is double and the other single occupied. In this case, $\Lambda = 1$; this state is also doubly degenerate with $+1$ and -1, respectively. The term symbol here is $^2\Pi$.

12.2.1 Oxygen Complexes

Three different coordination modes are observed for the binding of molecular oxygen to a metal center, each given with an example in Fig. 12.5. Besides the η^1 (end-on) and η^2 (side-on) coordination to a metal center, there is the bridging μ, $\eta^1 : \eta^1$ (bridging end-on) option as a third possibility.

For the discussion of bonding, oxygen is a σ-donor-π-acceptor ligand. For η^1 coordination, the σ bond is formed between an occupied molecular orbital of oxygen (which may be a σ or π orbital, similar to the CO ligand) and an empty d orbital. For the π backbonding, an occupied d orbital overlaps with an empty or half-occupied π^* orbital of oxygen. A very strong π backbonding can be equated with an intramolecular shift of electrons from the metal center to the oxygen. Therefore, the oxidation states from the oxygen molecule and the metal center in the complex cannot be determined beyond doubt. This finding, O_2 belongs to the "non-innocent" ligands, is important for further discussion. In side-on

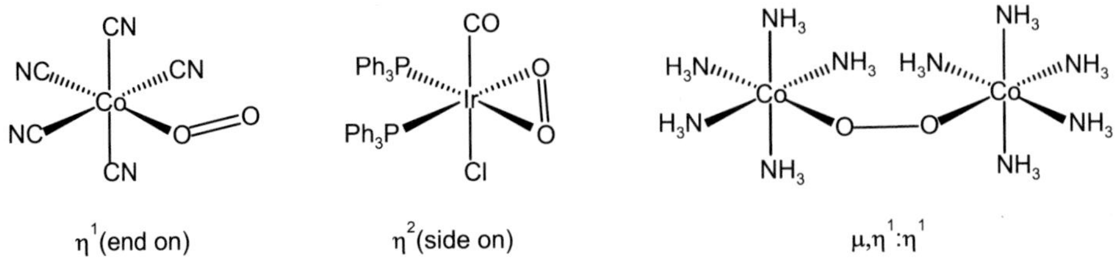

η^1(end on) η^2(side on) $\mu,\eta^1{:}\eta^1$

Fig. 12.5 Examples of complexes with molecular oxygen illustrating the different binding modes

coordination, the donor bond is a π bond between the partially filled π^* orbitals from the O_2 and empty d orbitals at the metal center. The back-bond between an occupied d orbital of the metal centre and an empty or half-occupied π^* orbital of the oxygen is formally a δ bond.

12.2.2 Bonding Situation in Hemoglobin

In higher organisms, the transport and storage of oxygen is realized by the heme proteins hemoglobin (transport) and myoglobin (storage). Other organisms (e.g. molluscs and crustaceans) use dinuclear iron or copper complexes for this function, in which the metal centers are coordinated directly to the protein backbone via amino acid side groups. Both hemoglobin and myoglobin are capable of reversibly binding oxygen. This commonality is reflected in the structure of the active site, which is the same for both proteins. A major difference between the two proteins is the number of heme subunits per protein. In hemoglobin there are four, whereas in myoglobin there is only one. The four subunits in hemoglobin are important for the cooperative effects of uptake (once one oxygen molecule is bound, the others are bound more rapidly) and release of oxygen, which will not be considered further in this book. We focus on the bonding between the oxygen and the iron center and how bond cleavage is prevented in the oxygen molecule. Bond cleavage is accompanied by a stepwise transfer of electrons to the oxygen, so a consideration of the oxidation states from the metal center and oxygen is particularly interesting.

  The crystal structure of hemoglobin provided the first insight into the bonding between iron and oxygen. The first single X-ray crystal structure analysis was carried out in the late 1950s; the first crystals of the red blood pigment were obtained much earlier, in 1849. In 1962, J. C. Kendrew and M. F. Perutz were awarded the Nobel Prize in Chemistry for the elucidation of the crystal structure of hemoglobin. In the oxygen-free *desoxy* form, the central iron(II) ion has a coordination number of five and is in a square pyramidal coordination environment. The iron is located slightly below the porphyrin macrocycle, which can be explained on the one hand by the square pyramidal coordination environment. In addition, the iron(II) ion is in the high-spin state ($S = 2$) and is thus slightly too large for the porphyrin ring. As can be seen in Fig. 12.6, the axial fifth ligand is the nitrogen atom from the imidazole five-membered ring of a histidine side chain, also known as the

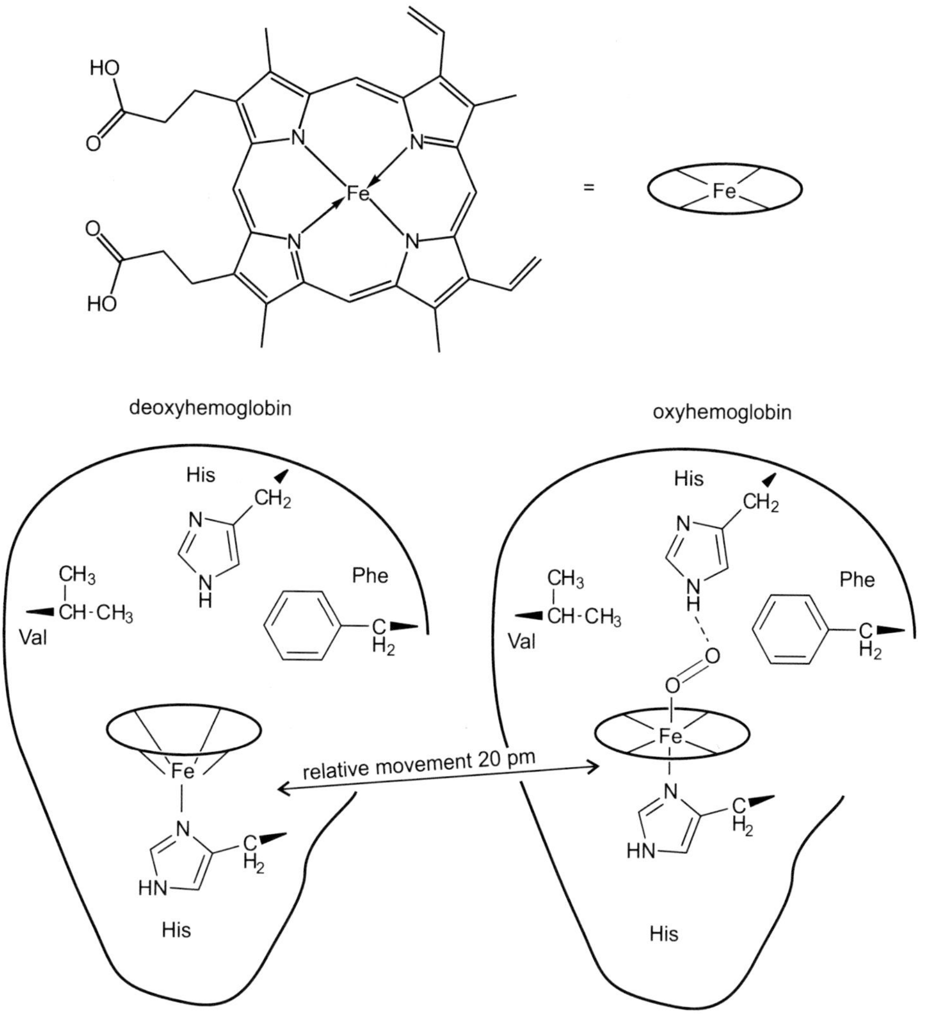

Fig. 12.6 Detail of the active site of hemoglobin/myoglobin in the deoxy and oxy forms. In the unloaded form, the iron ion has a coordination number of five and is located slightly below the porphyrin ring. When oxygen is coordinated, the coordination number increases from five to six and the iron ion is now in the plane of the porphyrin ring. This causes the axial histidine to move about 20 pm. This relative movement is important for the cooperative effects in hemoglobin

axial histidine. In the immediate vicinity from the active site are another histidine (referred to as distal histidine), a valine, and a phenylalanine side chain. The sixth coordination site is vacant and is occupied in the oxygen-loaded *oxy* form with oxygen as the sixth ligand. The oxygen thereby coordinates end-on, angled to the iron ion with an angle of about 120 °. With the coordination of the oxygen ($S = 1$), the iron now enters a low-spin state and the whole system is now diamagnetic ($S = 0$). We leave aside for the moment the question of the oxidation states of oxygen and iron. Due to the transition to the low-spin state, the diameter of the iron ion decreases and it now fits nicely into the porphyrin macrocycle, into which it now "slips", also due to the coordination number change. The iron now has coordination number six and lies exactly in the middle of the octahedron. This movement of the iron towards the porphyrin ring also changes the position of the axial histidine. In hemoglobin, this relative movement is responsible for the cooperative effects between the four heme subunits.

In the *oxy* form, the distal histidine forms a hydrogen bond to the coordinated oxygen. This hydrogen bonding and the shape of the protein pocket force the angled arrangement of the oxygen. In the discussion of the bonding situation in oxygen complexes, it has already been mentioned that a σ-donor-π-acceptor bond is formed between the metal center and the ligand. However, in the case of triplet oxygen, the π^* orbitals are both half-occupied and therefore not particularly suited for the formation of a π backbond. Much more suitable would be ligands with no (CO) or only one (NO) electron in the π^* orbitals, which can then form more stable complexes. Indeed, the complex formation constant for the coordination of CO to a protein-free heme is about 25,000 times larger than the complex formation constant for the coordination of oxygen. Under these conditions, even a small increase in carbon monoxide content in the atmosphere would be critical. The protein environment significantly increases the bonding selectivity for oxygen. The forced angled coordination makes the formation of π bonds more difficult. In addition, hydrogen bonds cannot be formed in the case of carbon monoxide. As a result, the complex formation constant for CO is only 200 times as large as that for O_2. Due to the considerably higher oxygen content in our atmosphere (20.9% vs. 50–200 ppb for CO), oxygen transport works in our organism. Too high a concentration of CO leads to intoxication, which can be treated by the administration of oxygen-enriched air.

The question of possible electron transfer between iron and oxygen during coordination from the oxygen to the heme center has long preoccupied scientists, and two alternative models have been discussed.

Formulation According to Pauling and Coryell

As early as 1936, it was proposed that in the *oxy* form there is a low-spin iron(II) center to which a singlet oxygen molecule is coordinated. The porphyrin macrocycle is a strong-field ligand, so that in an octahedral coordination environment the iron(II) ion is in the $S = 0$ low-spin state. Coordination of the oxygen to the iron removes the equivalence of the two π^* orbitals of the oxygen and the one lower in energy is occupied by both electrons, which are now paired. The oxygen is now in the $S = 0$ ($^1\Delta$) state. This formulation is supported by the high tendency of the heme center to form complexes with CO and NO, both of which prefer a low-valent metal center, i.e., iron in the +2 oxidation state.

Formulation According to Weiss

Approximately 30 years after Pauling and Coryell's proposal, an alternative formulation for describing the bonding situation in hemoglobin was proposed in 1964. Based on a series of spectroscopic results as well as further studies of reactivity, Weiss proposed in 1964 that the *oxy* form is better described as a low-spin iron(III) ion $(S = \frac{1}{2})$ with a superoxide radical anion $(O_2^{\cdot-}, S = \frac{1}{2})$. He explains the $S = 0$ total spin of the system by a strong antiferromagnetic coupling between the two unpaired electrons. The basis for this proposal is provided by the following findings.

- The O-O stretching vibration frequency, $\nu(\text{O-O}) = 1100 \text{ cm}^{-1}$, typical for a superoxide radical anion.
- In the Mössbauer spectrum, parameters are obtained that are characteristic of iron(III) in the low-spin state.
- The reactivity of the *oxy*-form is similar to that of analogous iron(III) pseudohalide complexes. Thus, the oxygen can be easily exchanged for chloride ions. Similar behaviour is observed with corresponding azide (N_3^-) complexes.
- If the iron in the active site is exchanged for a cobalt, then the corresponding cobalt(III)-oxygen complex has one electron more and is paramagnetic $(S = \frac{1}{2})$. Using ESR spectroscopy, it was shown that the unpaired electron in this complex resides predominantly at the oxygen, corresponding to the formulation Co(III) $(S = 0)$, $O_2^{\cdot-}$ $(S = \frac{1}{2})$.

These results suggest an electron transfer between the iron and the oxygen in the formation of the oxygen complex. However, it must be taken into account that the cobalt and the iron, for example, cannot be directly compared with each other as the active centre, due to the different electron configurations. Thus, it could well be that electron transfer occurs in the case of cobalt but not for iron. The difficulty in assigning oxidation states is due to the strongly covalent nature of the iron-oxygen bond. The formation of joined molecular orbitals leads to the fact that the electrons are delocalized over both bonding partners.

12.2.3   Model Compounds for Hemoglobin and Myoglobin

It has already been pointed out that heme, which is not protected by the protein environment, reacts irreversibly with oxygen to form a μ-oxido complex. Similar behavior is observed in many structural model compounds, an example of which is given in Fig. 12.3. This observation led to the question of the role of the protein environment for the reversible reaction with oxygen. In Fig. 12.7, the intermediate steps for the irreversible reaction are given. The reaction of the initial iron(II) complex with oxygen leads to the formation of a dimeric μ-peroxidoiron(III) complex. This is not stable and decomposes under bond cleavage into two iron oxido complexes in which the iron has the formal oxidation state +4. This highly reactive species, which is also postulated as an intermediate in catalytic cycles, e.g. in cytochrome P450, reacts immediately with another iron(II) starting complex to form the stable end product, the μ-oxido iron(III) complex.

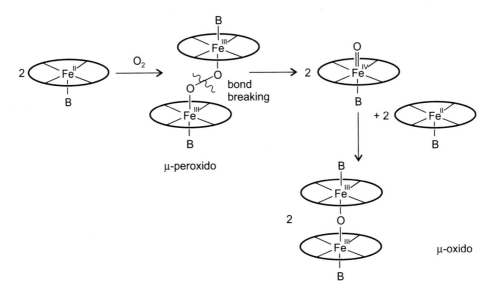

Fig. 12.7 Reaction of heme not protected by the protein environment with oxygen. B represents a base (e.g. imidazole or pyridine) that acts as a fifth ligand

The course of the reaction shows that for a reversible coordination of oxygen, the formation of a μ-peroxido complex and thus the oxygen-oxygen bond break must be prevented. This can be achieved with the help of picket fence porphyrins. By cleverly attaching sterically demanding substituents, one side of the iron center is shielded and dimerization is no longer possible. Examples of such model compounds are shown in Fig. 12.8 [88, 89]. The unsubstituted starting complex, iron tetraphenylporphyrin (FeTPP) with R = H, is often used as a structural model compound for the heme center. By introducing the substituents, the protein environment is simulated and the structural model compound approximates the model to the extent that the function of the natural model is also reproduced. This process of continuous adaptation of model compounds helps us to better understand the function of the biological model. In further optimization steps, systems were developed in which the axial ligand (B in Fig. 12.7) is covalently bound to the porphyrin ring. There have also been initial attempts to simulate the phenomenon of cooperativity in hemoglobin in model compounds [89].

The first synthetic model compound for hemoglobin or myoglobin, which reflected biological function, was prepared as early as 1938. It is cobalt salen; in Fig. 12.2 the structure of the salen ligand is shown on the right. Other group 9 complexes (Co, Rh, Ir) also show the possibility of reversible binding of oxygen. An example is the Vaska's complex [IrCl(CO)(PPh$_3$)$_2$], which is square planar (the iridium(I) ion is a d^8 system). Reaction with oxygen leads to an octahedral iridium(III) complex (d^6 system) with oxygen bound side-on as shown in the middle of Fig. 12.5. In cobalt salen, oxygen uptake occurs at room temperature. The powder, which is red in the oxygen-free form, turns black and the

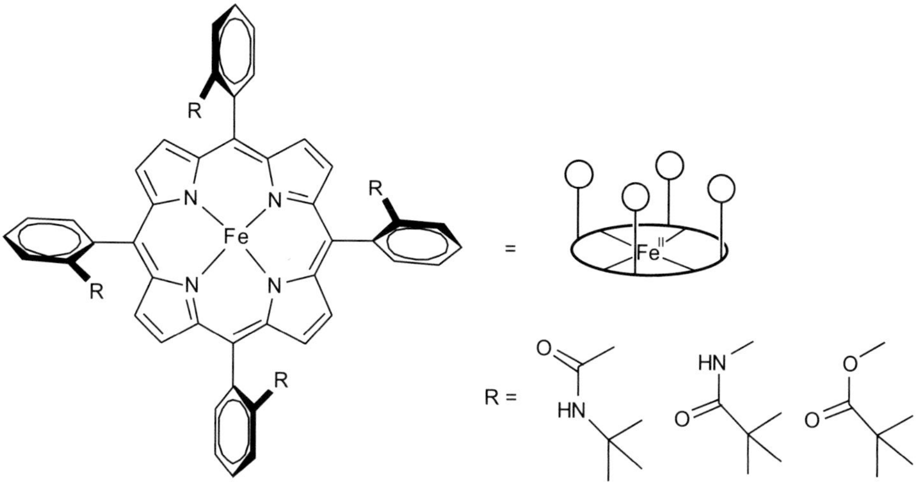

Fig. 12.8 Examples of picket-fence porphyrins. All examples start from the tetraphenylporphyrin (TPP), where R = H and which is frequently used as a model compound for the heme. By introducing sterically demanding substituents R in the *ortho-position* from the porphyrin ring, the iron center is shielded on one side. If the fifth ligand (B in Fig. 12.7) coordinates at the unshielded side, then oxygen coordinates in a shielded pocket and dimerization is prevented

change in weight speaks for the uptake of one oxygen molecule per cobalt atom. At temperatures above 100 °C, the oxygen is released again and the original cobalt (II) complex is reformed. The corresponding iron(II) salen complex reacts irreversibly with oxygen to form the μ-oxido iron(III) complex. The reason for the different reactivity of iron and cobalt can be explained by their electron configuration and the resulting preferred oxidation states. In the case of iron, the preferred oxidation state is +3, in which the 3d shell is half occupied and which is therefore particularly stable. The +2 oxidation state is only preferred in the presence of strong-field ligands with an octahedral coordination environment (high ligand field stabilization energy for a d^6 system in the low-spin state). For cobalt, the situation is the other way around. In the aqueous medium, cobalt is usually present in the +2 oxidation state, where only the two 4s electrons have been removed. The removal of a further electron is only possible in the presence of relatively strong ligands with the formation of octahedral complexes (same explanation as for iron(II)). In the case of cobalt, this reversed reactivity favours the reversible binding of oxygen. The fact that nature has chosen iron as the metal centre is due to its significantly better bioavailability.

12.3   Cobalamins: Stable Organometallic Compounds

Cobalamins are metalloproteins containing the trace element cobalt as a central atom. The vitamins B_{12} and the coenzyme B_{12} derived from them are important for humans. Figure 12.9 shows the general structure of the cobalamins. They differ in the axial ligand R. In the case of vitamins B_{12}, R = CN, OH or H_2O, we then also speak of cyanocobalamin (Vit B_{12}), hydroxocobalamin (Vit B_{12b}) or aquacobalamin (Vit B_{12a}). The biologically active forms are methylcobalamin (R = Me, MeB$_{12}$) and 5'-deoxyadenosylcobalamin (R = 5'-deoxyadenosyl, the coenzyme B_{12}). Both compounds have a cobalt-carbon bond. They are classic organometallic compounds which are stable under physiological conditions.

Coenzyme B_{12} is a cofactor. This means that it is essential for the function of certain enzymes, but does not participate directly in the reaction. What this means can be well illustrated by the example of the coenzyme B_{12}. Its function is the controlled formation of free radicals. These radicals are then required by the enzyme for the reaction to be catalyzed, e.g. for a 1,2-shift. The enzyme is responsible for the binding of the substrate (substrate specificity), here the actual reaction takes place at the active site. The coenzyme is not directly involved in this reaction. An enzyme without a coenzyme is also called an

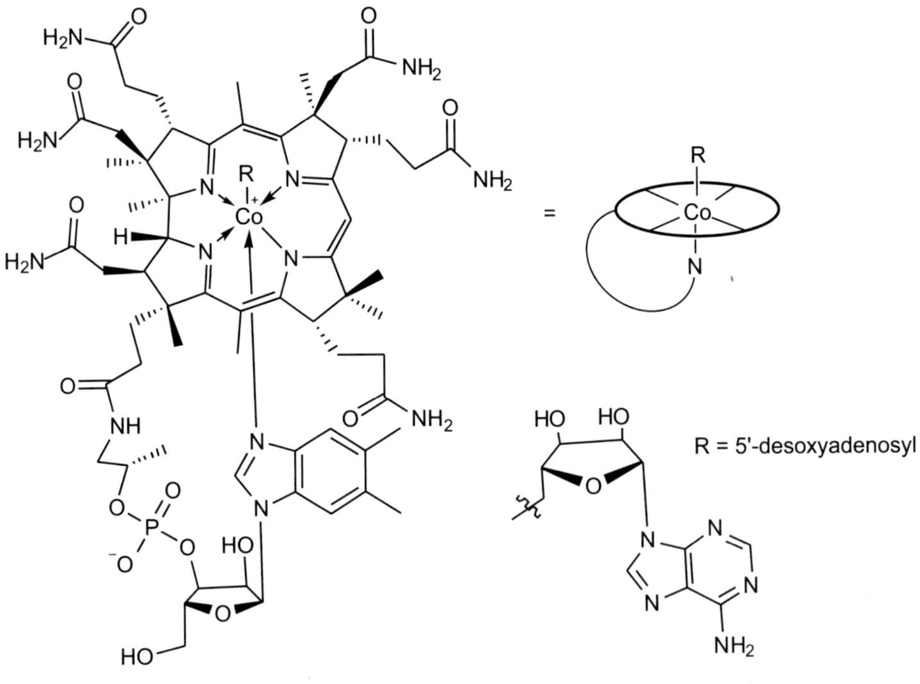

Fig. 12.9 General structure of cobalamins. For the biologically inactive forms, R = CN, OH or H_2O, they are known as vitamins B_{12}. The biologically active forms are organometallic compounds, with R = Me or 5'-deoxyadenosyl. The latter variant is the coenzyme B_{12}, which is responsible for the controlled formation of free radicals in the human body

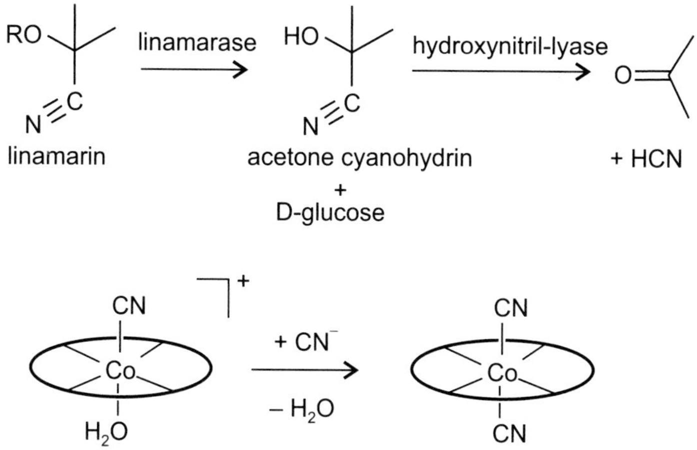

Fig. 12.10 The detection of biological cyanide is based on a rapid color change of cobalamin from yellow to red in the presence of cyanide, which is due to a ligand exchange

apoenzyme, while the working unit of apoenzyme and coenzyme, the total enzyme, is also called a holoenzyme.

The interest in cobalamins is not limited to bioinorganic issues. The coenzyme B_{12} or model compounds thereof are partly used in organic chemistry for the realization of 1,2-shifts. An emerging field of work is the detection of biological cyanide [90]. Some species of cassava (a plant used as a starchy staple) contain cyanogenic glucosides such as linamarin (see Fig. 12.10). These cyanogenic glycosides can be hydrolyzed by the plant's own enzymes (e.g., linamarase). One of the products is acetone cyanohydrin, which further decomposes to acetone and hydrogen cyanide. In addition to cassava, similar endogenous cyanides are found in bamboo and flax, which are staple foods for many people in Africa, Southeast Asia, and Latin America. High levels of biological cyanides lead to chronic cyanide poisoning in the long term. To rapidly detect biological cyanide, derivatives of cobalamines can be used in which a water molecule is exchanged for cyanide as an axial ligand. The reaction equation is given in Fig. 12.10 below. The ligand exchange results in a color change that is readily visible to the naked eye. That ligand exchange is accompanied by a color change is not unexpected (See Chap. 5, Color of Complexes). The interesting thing about this reaction is the ligand exchange itself, which is very fast. In Sect. 6.2 (stability of complexes) we learned that octahedral cobalt(III) complexes have a very high kinetic stability and therefore ligand exchange reactions are very slow.

In the following we will deal with how the free radicals are formed, which special role cobalt plays in this process and what enables the rapid ligand exchange in cobalamins.

12.3.1 Bioavailability of Elements

The first functional model compounds for the reversible binding of oxygen were cobalt (II) complexes. In biological systems, iron and copper complexes are responsible for this functionality. At first glance, this is unexpected. From a coordination chemistry point of view, cobalt and its higher homologues rhodium and iridium are much better suited for this function. To realize the same reactivity with iron, further general conditions (see picket-fence porphyrins) have to be considered. Reason for this is the different bioavailability of the individual elements. Table 12.2 gives the abundance of selected biologically relevant elements. For bioavailability, the content in seawater is of particular interest, since evolution began here. For some elements, bioavailability has changed in the course of evolution. A good example is iron, which was present in divalent form before the formation of the oxygen atmosphere and was then readily bioavailable (because it was readily soluble in water). In the +3 oxidation state, the appearance of poorly soluble oxides and hydroxides leads to a much lower iron content in seawater. A look at cobalt shows that it has by far the lowest bioavailability of the 3d elements. In general, cobalt is the rarest element of the 3d series. In addition, cobalt has only one function in the body – that of coenzyme B_{12}. This is

Table 12.2 Frequency of selected biologically relevant elements. The elements of the 3d series are highlighted [86]

Element	Human body [mg/kg]	See water [mg/l]	Earth crust [mg/kg]	Distribution [wt %]
H	101,000	107,300	1,400	7.7
O	654,000	860,500	467,600	48.9
C	181,000	28	200	0.02
N	30,000	0.5	20	0.017
Ca	15,000	410	36,400	3.4
P	10,000	0.07	1,000	0.1
S	2,500	928	260	0.03
K	2,200	380	26,000	2.4
Na	1,500	11,000	28,400	2.7
Cl	1,400	18,100	130	0.11
Mg	470	1,300	21,000	2.0
Fe	60	3×10^{-3}	50,200	4.7
Zn	40	5×10^{-3}	70	7×10^{-3}
Si	20	1	278,600	26.3
F	10	1.4	625	0.06
Sr	4	8.5	375	0.036
Cu	3	3×10^{-3}	55	5×10^{-3}
I	1	0.06	0.5	5×10^{-5}
Mn	0.3	2×10^{-3}	950	0.091
V	0.3	1.5×10^{-3}	135	0.013
Se	0.2	4.5×10^{-4}	0.05	5×10^{-6}
Mo	0.07	0.01	1.5	1.4×10^{-4}
Cr	0.03	6×10^{-4}	100	0.01
Co	0.03	8×10^{-5}	25	4×10^{-3}
Ni	0.014	2×10^{-3}	75	7.2×10^{-3}

different from iron, which has a large number of different functions in the human body (see Table 12.1). This circumstance indicates that only cobalt, in combination with the corrin ligand, is suitable for the specific reactivity in coenzyme B_{12}.

12.3.2 Structure

The discovery of coenzyme B_{12} took place much later than that of the red blood pigment hemoglobin, which was isolated in crystalline form as early as the mid-nineteenth century. The reason for this are the different concentrations in which the two components are present in the blood. In the case of cobalamins, it is about 0.01 mg/l blood, which made enrichment and isolation of the compound much more difficult. This was only achieved with the realization of chromatographic separation methods, and thus cyanocobalamin could be prepared pure for the first time in 1948. The existence of an "essential component" was discovered as early as the 1920s. At that time, severe forms of anemia were treated by the administration of liver extracts. In these extracts, an "essential component" was detected, which is cobalt-containing and can only be synthesized by microorganisms. Essential (organic) compounds that cannot be produced by the body are called vitamins. For this reason, the essential component from liver extracts was called vitamin B_{12}. In 1964 Dorothy Crowfoot-Hodgkin succeeded in the crystal structure analysis of vitamin B_{12} and later of the coenzyme. With approximately 100 non-hydrogen atoms, this was an outstanding achievement at the time, which made an important contribution to the understanding of cobalamins and for which Crowfoot-Hodgkin was awarded the Nobel Prize in Chemistry.

Figure 12.9 shows the general structure of the cobalamins, which differ only in the axial ligand R. As with heme, the cobalt is located in the middle of a macrocyclic N_4 ligand. However, this is the corrin ligand, which differs from the porphyrins in having a smaller ring size (15 membered instead of 16 membered) and a single negative charge (instead of a double negative charge). Cobalt-porphyrin complexes can be prepared and are also stable, but they show a different reactivity and are not suitable as cobalamin substitutes. In the fifth position, the cobalt has another axial ligand. This is a 5,6-dimethylbenzimidazole ring coordinated via N(1), which is connected to the corrin macrocycle via a longer chain. The corrin macrocycle, unlike the porphyrin macrocycle, is not planar but folded in a butterfly or saddle conformation. Modeling studies show that this distortion is important for cobalamin reactivity. As with the porphyrins, the corrin ligand is also a pronounced strong-field ligand and low-spin complexes are usually obtained, although they have a distorted structure. This affects the splitting of the d-orbitals of the cobalt and thus also the reactivity of the compound. Thus, the distorted structure is certainly one of the reasons why the cobalamins are capable of a comparatively fast ligand exchange.

12.3.3 Reactivity

In coenzyme B_{12} the cobalt has the oxidation state +3 and the coordination number 6. The initial situation is comparable to Werner's complexes (Chap. 1). The strong-field ligand and the high positive charge of the metal ion lead to a strong splitting of the d orbitals in the octahedral ligand field. The six d electrons of the cobalt(III) ion are paired in lower t_{2g} orbitals and the complex is diamagnetic. The high stability of octahedral cobalt(III) complexes is related to the high ligand field stabilization energy.

The reactivity of coenzyme B_{12} is accompanied by breaking the cobalt-carbon bond. Three possibilities are conceivable here, which are given in Fig. 12.11. The cobalt-carbon bond can be cleaved homolytically or heterolytically. In homolytic bond cleavage, the two cleavage products are each assigned one electron from the common electron pair. In the case of coenzyme B_{12}, this results in the formation of a cobalt(II) complex and an organic radical. The cobalt(II) complex now has a square pyramidal coordination environment. This coordination environment is more favorable for a cobalt(II) ion with seven d electrons than the octahedral coordination environment because the splitting of the e_g orbitals allows the now lower-lying d_{z^2} orbital to be occupied by the additional electron, thereby obtaining a higher ligand field stabilization energy. In heterolytic bond cleavage, the shared electron pair is assigned to one of the two bonding partners. The two possibilities that occur here are shown on the left and right in Fig. 12.11. If the shared electron pair remains at substituent R, then a carbanion and a cobalt(III) complex are formally formed. This variant of heterolytic bond cleavage plays a role in ligand exchange reactions in which the oxidation state of the metal center and the coordination number are preserved. An example of this would be the formation of coenzyme B_{12} from vitamin B_{12}.

In the variant shown on the right in Fig. 12.11, the shared electron pair remains on the cobalt and a cobalt(I) complex and a carbocation are obtained. It should be emphasized that in this case the fifth ligand moves completely away from the cobalt and the latter is now in a square planar coordination environment. This is particularly favorable for the cobalt(I) ion with its eight d electrons, as again a high ligand field stabilization energy is realized. The strongly antibonding $d_{x^2-y^2}$ orbital in this coordination geometry is not occupied, resulting in an energy gain for the overall system. The redox potentials for the different steps show that homolysis is in the physiologically interesting range. This is the preferred reaction.

The redox potentials of the two reduction steps from cobalt(III) to cobalt(II) and cobalt (I) are influenced by the axial ligand. The decrease of the oxidation state is accompanied by a decrease of the coordination number until the fifth ligand is completely cleaved. For this reason, this ligand is also referred to as the control ligand. The step from the octahedral cobalt(III) complex to the square planar cobalt(I) complex corresponds to the reductive elimination known from organometallic chemistry (the reverse reaction is oxidative addition).

With the corrin ligand, the cobalt(I) step is stable even under physiological conditions. This is not the case with the analogous cobalt porphyrin complexes. Reasons for this could

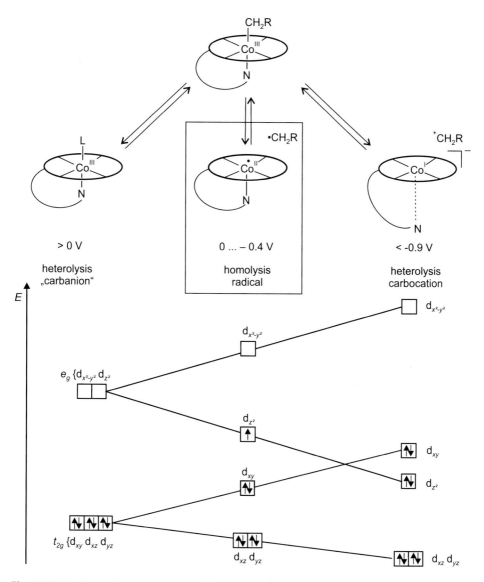

Fig. 12.11 In the resting state, coenzyme B_{12} is a cobalt(III) complex with an octahedral ligand field that exists in the low-spin state. The cobalt-carbon bond can be cleaved homolytically or heterolytically. The former leads to the formation of a square pyramidal cobalt(II) complex and a radical (middle). In heterolytic bond cleavage, the shared electron pair remains with one of the two bond partners. For the variant shown on the left, a carbanion is formally generated and the shared electron pair remains with the carbon. This reaction corresponds to a ligand exchange, as occurs, for example, in the generation of coenzyme B_{12} from vitamin B_{12}. In the variant shown on the right, the shared electron pair remains with the cobalt, and a carbocation and a cobalt(I) complex are formed. The different oxidation states are stabilized by the different coordination geometries. The position of the fifth ligand, which is also referred to as the control ligand, plays an important role here

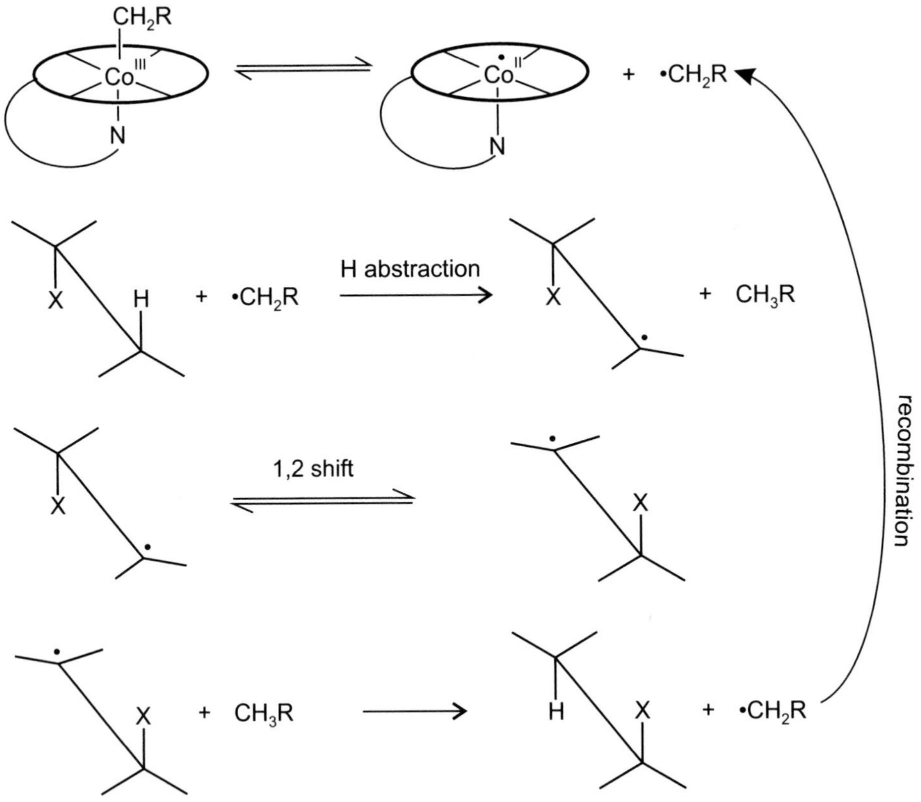

Fig. 12.12 Mutase activity of coenzyme B_{12}; the coenzyme B_{12} provides the radical required for the 1,2-shift

be a somewhat weaker ligand field of the porphyrin ligand or also the onefold negative charge of the corrin ligand.

Mutase Activity

It was already mentioned at the beginning that the controlled generation of radicals is one function of the coenzyme B_{12}. The radicals generated in this process are required, for example, for mutase reactions. The general equation for this is given in Fig. 12.12. We see that the cobalt is not directly involved in the reaction, but only provides the radical needed for it. The role of the enzyme is substrate binding. This triggers a conformational change that leads to a decrease in the cobalt-carbon binding energy and initiates reversible homolysis of the cobalt-carbon bond. In addition, the protein environment of the enzyme shields the radical to avoid unwanted reactions and is responsible for the stereoselectivity of the reaction. An example is glutamate mutase, which catalyzes the conversion of glutamic acid to β-methylaspartic acid (Fig. 12.13 above).

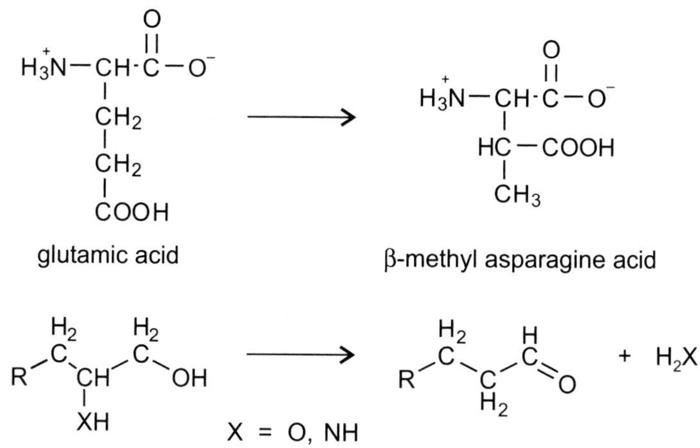

Fig. 12.13 Top: Mutase-directed conversion of glutamic acid to β-methylaspartic acid. Bottom: Coenzyme B$_{12}$-dependent 1,2-shift of HX leading to subsequent elimination, as observed in dehydratases and deaminases

The 1,2-shift of functional groups HX, which can be OH- or NH$_2$-groups, is important for dehydratases and deaminases, which are also coenzyme B$_{12}$-dependent. Here, the 1,2-shift leads to an unstable intermediate followed by double bond formation with cleavage of H$_2$X. The general mechanism is shown in Fig. 12.13, bottom.

Alkylation Reaction

In microorganisms, methyl group-transferring reactions are important in addition to the reactions discussed so far. An example that is also relevant for us is the synthesis of the essential amino acid methionine from homocysteine, which is carried out, for example, by the methionine synthase in *E. coli*. In this process, methylcobalamin acts as a methylating agent. The methyl group source is, for example, 5-methyltetrahydrofolic acid (5-methyl-THFA). The corresponding equation is given in Fig. 12.14. In this reaction, an electrophilic methyl group is transferred, that is, a carbocation and the corresponding cobalt(I) species are formed (see Fig. 12.11). Depending on the substrate to which the methyl group is to be transferred, the methylation reactions may also proceed radically or via the formation of a carbanion. The latter is thought to occur for noble elements such as mercury. Methylation leads to the formation of methylmercury (MeHg$^+$), a particularly toxic form of mercury. The high toxicity can be explained by the at the same time lipophilic and hydrophilic character of the small molecule. It allows the mercury to penetrate the blood-brain barrier or the placental membrane and thus to be distributed almost unhindered in the body.

Model Compounds and Application in Organic Chemistry

1,2-shifts look simple at first sight, but are difficult to perform in organic chemistry. For this reason, there is great interest in cobalamines and corresponding model compounds. For use

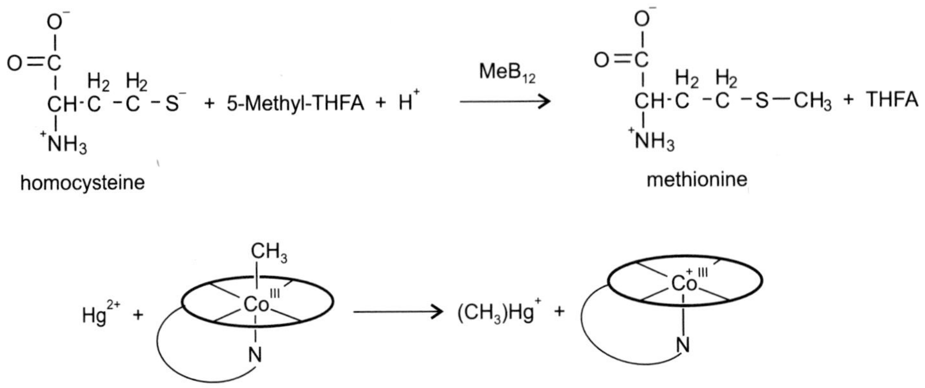

Fig. 12.14 Methylcobalamin-mediated methylation reactions can proceed via a carbocation, radical or via a carbanion. An example of the first variant is the synthesis of the essential amino acid methionine shown above. A carbanionic mechanism is assumed for the formation of methylmercury

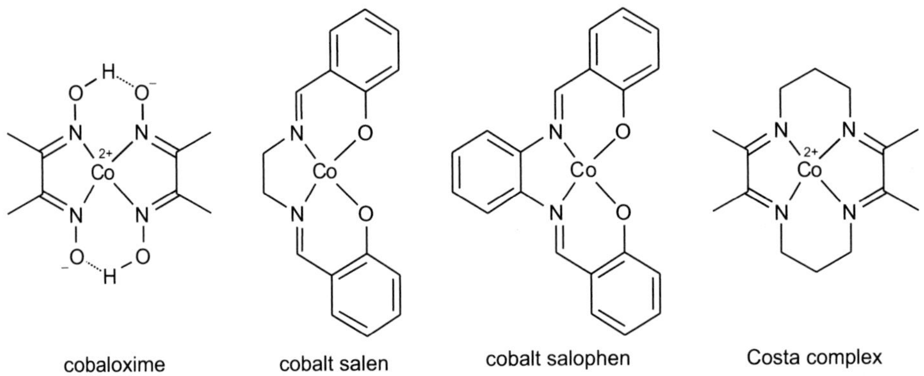

cobaloxime   cobalt salen   cobalt salophen   Costa complex

Fig. 12.15 Model compounds for cobalamins

in organic synthesis, model compounds are expected to be less sensitive and more soluble, especially in organic solvents, than cobalamines. So far, mainly structural model compounds have been prepared, which mimic the function of the cobalamines only to a very limited extent. Examples of these are given in Fig. 12.15. The cobaloxime (the bis (diacetyldioximato)cobalt(II)) was the first model compound to reflect well some of the properties of the cobalamins. Similar properties are exhibited by the cobalt-salen and -salophene complexes. The Costa complex is the only example with a macrocyclic ligand. Here, good agreement was found with the redox potentials of the original. Similarly to the cobalt porphyrins, all model compounds fail with regard to the stability of the supernucleophilic cobalt(I) step. Alternatively, cobester complexes were prepared. These are B_{12} derivatives without a nucleotide side chain, prepared from cobalamines.

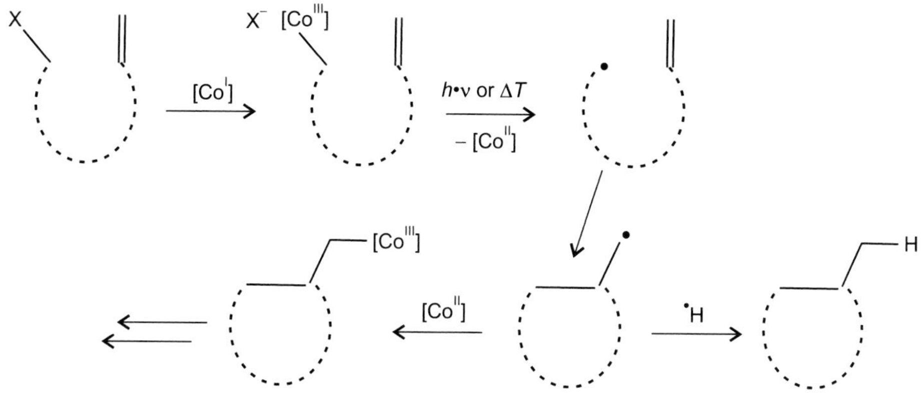

Fig. 12.16 Cobalt complex-assisted ring closure

The ability to form alkyl-cobalt bonds and their easy cleavage under the formation of radicals make vitamin B_{12} and analogous cobalt complexes an interesting tool in organic synthesis and natural product synthesis [91]. Usually, these reactions involve C-C bond formation, and the general mechanism is given in Fig. 12.16 using ring closure reactions as an example. The first step is the substitution of a halogen X by the supernucleophilic cobalt (I) complex to form a cobalt-carbon bond. This can be cleaved by irradiation or temperature elevation to form an organic radical and a cobalt(II) species. This is followed by ring closure. The resulting radical can be captured or recombined with the still present cobalt (II) complex. This variant opens up the possibility of further functionalization steps.

12.4   Questions

- What is the difference between a structural and a functional model compound?
- What is meant by a donor bond or acceptor bond? Draw the relevant molecular orbitals of a corresponding complex with molecular oxygen in the end-on or side-on coordination. In the case of the metal centre and the oxygen molecule, label which orbitals overlap!
- What happens during carbon monoxide poisoning and why does the administration of oxygen-enriched air help treat it?
- Why is a square planar ligand field required for the stabilization of the cobalt(I) step?

Catalysis

13

The importance of complexes or organometallic compounds for catalysis has already been reflected in the brief historical outline of the development of organometallic chemistry. This is underlined by the large number of Nobel Prizes awarded for catalytic processes. In the following, the example of polymerization catalysis and its discovery will be discussed in the first section. The book "Organometallic Chemistry" by Elschenbroich [15] is recommended as further reading on this current field of research. The second section of this chapter deals with photocatalysis, which is highly relevant, especially for meeting the challenges of today. For example, the book "Chemical Photocatalysis" by König (Ed.) [92] is recommended as further reading.

13.1 Catalyst

Before we look at catalytic processes, the term catalyst should first be explained. IUPAC says: A catalyst is a substance that increases the rate of reaction without changing the overall standard Gibbs energy change ΔG^0 of the reaction. The process is called catalysis. That is, a catalyst speeds up a reaction without being consumed. It is present unchanged after the reaction and therefore often does not appear in the reaction equation. In addition to the reaction rate, the catalyst often influences the reaction mechanism, usually by lowering the activation energy E_A for a reaction. However, it does not influence the position of the equilibrium. This relationship is illustrated in the energy diagram in Fig. 13.1.

The reaction equations given in Fig. 13.1 show that a product is formed intermediately from catalyst and reactant, which then decomposes again with recovery of the catalyst. In this connection, two other terms will be defined which will be needed in the following.

© The Author(s), under exclusive license to Springer-Verlag GmbH, DE, part of Springer Nature 2023
B. Weber, *Coordination Chemistry*,
https://doi.org/10.1007/978-3-662-66441-4_13

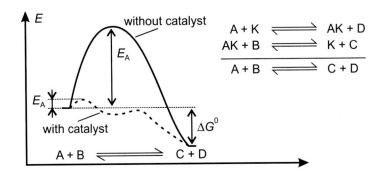

Fig. 13.1 Schematic representation of the influence of a catalyst on the energy profile of a reaction *(left)* and the course of the reaction *(right)*. The dashed curve with catalyst shows two maxima. This indicates that the reaction mechanism (see equations) has changed here and that the activation energy is lowered as a result

When speaking of catalytic processes, a distinction is made between homogeneous and heterogeneous catalysis. In homogeneous catalysis, the catalyst and reactants are present in one phase. This would be the case if everything is dissolved in the same solvent, for example, and everything is in a liquid phase. A gas phase reaction where reactants and the catalyst are in the gas phase is also homogeneous catalysis. Heterogeneous catalysis is when the catalyst and reactants are in different phases. A typical example is a solid catalyst where the reactants are liquid or gas and the catalyst is dispersed in this liquid or gas phase. In this case, the reaction takes place on the surface of the solid catalyst.

13.2   Preparation of Polyethylene (PE)

Polymerization catalysis is a catalytic process for producing long-chain polymers from simple olefins (e.g. ethene, also referred to as ethylene in the following). The resulting products – e.g. polyethylene (PE) – are important materials without which we cannot even imagine our lives. Even before the discovery of the transition metal-catalyzed polymerization of ethylene, which is of interest to us in this book, PE was being produced industrially. In the process known as the high-pressure process, ethene was polymerized radically at 1000–2000 bar and temperatures up to 300 $^\circ$C with oxygen as the initiator. Due to the extremely high pressures and high temperatures, high demands are placed on the reactors required for this process. Figure 13.2 briefly summarizes the reactions that take place. Under the reaction conditions, radicals are formed from the ethene (initiation), to which further ethene molecules attach themselves. This process is called chain growth or propagation. A break in chain growth occurs when two radicals "meet" – a process known as recombination, or when a hydrogen is transferred from one radical to a second, yielding an alkane and an alkene. This process corresponds to disproportionation. Another important

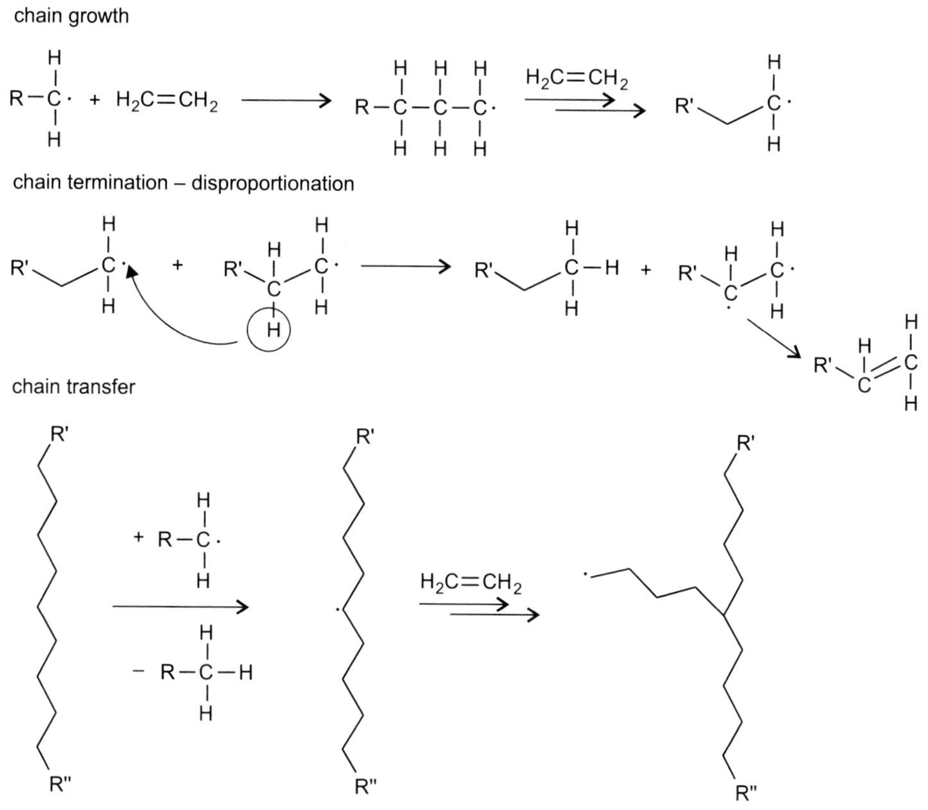

chain growth

chain termination – disproportionation

chain transfer

Fig. 13.2 Radical polymerization of ethene. Shown is the chain growth, the disproportionation as an example of a termination reaction and the chain transfer responsible for the branching of the resulting polymers

reaction is chain transfer. In it, a radical abstracts a hydrogen from the middle of an alkyl chain. A new radical is formed, on which a new chain grows, which corresponds to a branching.

13.2.1 Ziegler's Build-Up and Displacement Reaction

The discovery of transition metal-catalyzed polymerization catalysis began with studies by Karl Ziegler in the 1950s on the "build-up reaction". He investigated the reaction of triethylaluminum with ethylene under high temperatures and pressures and observed an addition of the Al-C bond to the $C = C$ double bond of ethene, i.e. an insertion of ethylene into the metal-carbon bond. In this way, he succeeded in synthesizing trialkylaluminum compounds whose alkyl chains could reach molecular weights of up to 3000 (up to 100 insertions). Chain growth cannot be continued at will. Above a certain chain length,

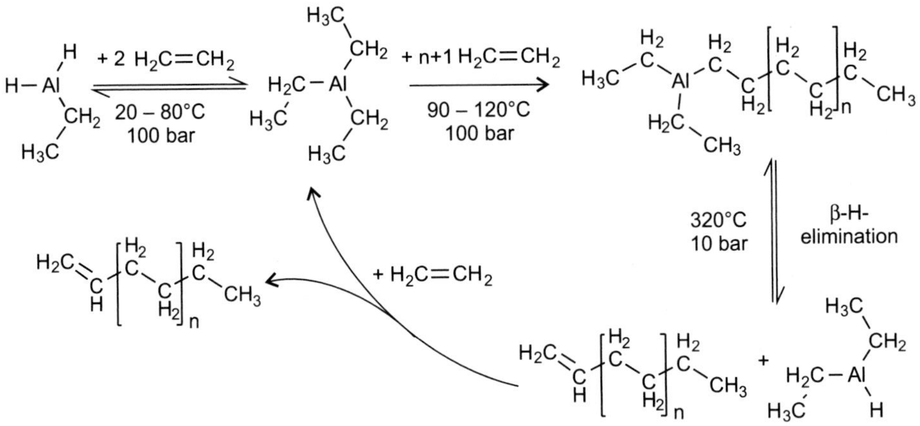

Fig. 13.3 Sequence of Ziegler's build-up reaction and displacement reaction. The different reaction conditions allow a two-step process in which both steps occur separately. At intermediate temperatures, both steps occur in parallel and catalytic oligomerization of ethene is possible. For clarity, the build-up reaction and displacement reaction have been shown for only one of the ethyl groups of aluminum. In fact, this process takes place on all three chains

β-H-elimination takes place. The β-H elimination is the reverse reaction of the olefin insertion. Both reactions are in equilibrium with each other (see Chap. 3, Elementary Reactions of Organometallic Chemistry) and the position of the equilibrium can be influenced to a certain extent by the reaction conditions (e.g. the ethylene pressure). The olefin formed in the β-H elimination is displaced by the ethene present in excess. This step is also known as Ziegler's displacement reaction. Figure 13.3 shows the reactions in equilibrium with each other. We can see that this is an ethylene oligomerization catalyzed by triethylaluminum. For polymerization to occur, over 1000 insertions would have to take place, but this does not succeed under the reaction conditions described here.

Like radical polymerization, Ziegler's build-up reaction for the oligomerization of ethene can be divided into chain start (insertion of ethene into the Al-H bond starting from ethylaluminum hydride), chain growth (the build-up reaction) and chain termination (β-H elimination). In the displacement reaction, the catalyst is regenerated and a catalytic process of ethene oligomerization is possible. It must be taken into account that the build-up reaction proceeds optimally at temperatures around 100 °C and high ethene pressures (100 bar), while higher temperatures (above 300 °C) and a lower ethene excess clearly favour the β-H elimination and thus the displacement reaction. Therefore, it is possible to carry out the oligomerization with the reaction conditions listed in Fig. 13.3 as a two-step process. The length of the olefins obtained can be controlled by the ratio of triethylaluminum to ethene. At temperatures between 180 °C and 200 °C, a catalytic process is possible in which shorter olefins (C4–C30 chain length) are obtained. This variant is still used today as a large-scale process (*Gulf process*, 200 °C, 250 bar).

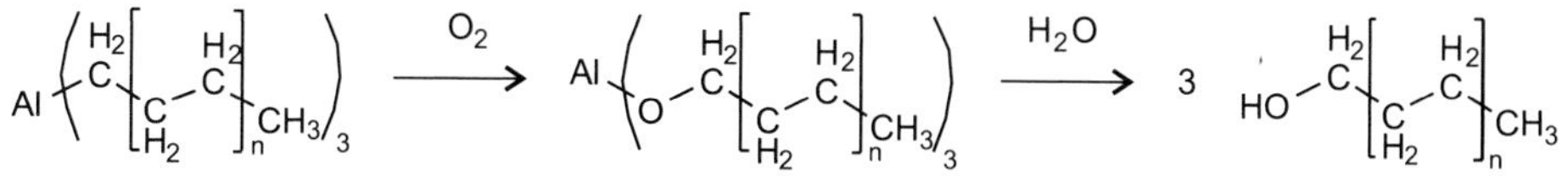

Fig. 13.4 Synthesis of fatty alcohols starting from trialkylaluminium compounds for the preparation of biodegradable detergents

With the conditions described so far, longer-chain olefins can be produced. Another of Ziegler's achievements is the Ziegler alcohol process, which is known as the *Alfol process* as a large-scale process. In this two-step process, the trialkylaluminum compounds formed in the buildup reaction are oxidized with oxygen to aluminum alcoholates in a second step and then hydrolyzed (Fig. 13.4). Synthetic fatty alcohols are formed in which chain lengths are set between 12 and 16 C atoms. These fatty alcohols are ideal starting materials for the production of biodegradable detergents. The process has thus made a significant contribution to reducing the detergent load in rivers and lakes.

Now we can ask ourselves why the aluminium ion in particular is suitable for the Ziegler build-up reaction. The build-up reaction corresponds to an insertion of olefins into a metal-carbon bond. This reaction requires a free coordination site at the metal center where the olefin can coordinate. In addition, a metal-carbon bond is also required. Both of these requirements are met in the case of triethylaluminum. On the one hand, an aluminum-carbon bond is present, and on the other hand, a free coordination site is still present, since the aluminum(III) ion strives for coordination number 4. The triethylaluminum is a Lewis acid, which lacks one pair of electrons to reach the noble gas configuration (in this case 8 valence electrons). For example, trimethylaluminum is dimeric in solid and dissolved phases (hydrocarbons) at room temperature and dissociates only in the gas phase at higher temperatures.

13.2.2 The Nickel Effect

Starting point for the discovery of the Ziegler-Natta catalysts for the polymerization of ethylene in the low-pressure process were studies by Karl Ziegler on the build-up reaction presented in the previous section. In one of these reactions, 1-butene was selectively produced instead of long-chain olefins. Subsequent investigations revealed that traces of a nickel compound were present in the reaction vessel. This observation, known as the nickel effect, led to the systematic investigation into the influence of heavy metals on Ziegler's build-up reaction [93]. The coordination chemist is now faced with the question of why short-chain olefins are preferentially obtained in the presence of nickel, whereas Ziegler-Natta catalysts (with titanium) lead to the formation of polymers. The decisive factor here is the different valence electron number of the active transition metal center. In the presence of triethylaluminum, the nickel, regardless of which oxidation state it started with, is reduced to nickel(0), which has a d^{10} electron configuration. Due to the large number of d electrons, nickel(0) is particularly suitable for the formation of complexes with olefins (strong π-backbonding). The high stability of nickel-olefin complexes favors β-H elimination (responsible for termination of chain growth) and only short olefins are obtained. One could speak of a nickel-catalyzed displacement reaction, which now proceeds much faster. The nickel(0) acts as a co-catalyst. It should be noted that nickel(0) is in itself also a catalyst with which, for example, butadiene can be cyclo-oligomerized to cyclooctadiene and cyclododecatriene.

To understand the nickel-catalyzed displacement reaction, one must consider the tendency of the Lewis acid trialkylaluminum to form Lewis acide-base adducts and multicenter bonds in the presence of a Lewis base (e.g., nickel(0)). Based on this, the following mechanism for the "nickel effect" was proposed (Fig. 13.5). The driving force for the reaction is the formation of butene, which is energetically more favorable than ethene.

1. The nickel(II) compound used as catalyst is reduced by trialkylaluminum.
2. Nickel(0) reacts with the ethene present in the reaction mixture to form tris-(ethylene)-nickel(0).
3. Between the nickel(0) complex and the trialkylaluminum, the multicenter bond shown in Fig. 13.5 is formed, in which the α -C atoms of the trialkylaluminum serves as a bridge between nickel and aluminum.
4. The bonds are rearranged in the sense of an electrocyclic reaction.

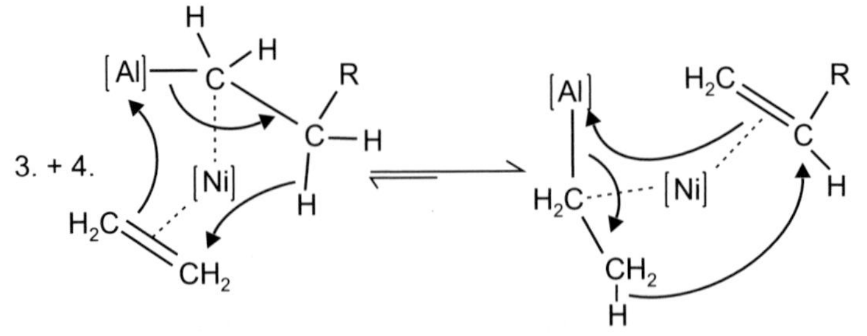

Fig. 13.5 Interpretation of the "nickel effect". A multicentre bond is formed between the tris (ethylene)nickel(0) and the trialkylaluminum. The α -C atoms of the trialkylaluminum coordinate to the nickel. Due to the spatial proximity, a rearrangement of the bonds in the sense of an electrocyclic reaction can occur

13.2.3 Polymerisation of Ethylene by the Low Pressure Process

The discovery of the nickel effect triggered systematic investigations into the influence of heavy metals (these are metals with a density greater than 5 g/cm^3) on the build-up and displacement reaction. Cobalt and platinum were also observed to accelerate the displacement reaction, while other metals showed no influence. Zirconium was the first metal for which polymerization of ethylene was observed. Further studies with transition metals of the fourth, fifth and sixth group showed that the most effective catalysts were obtained with titanium compounds. A catalyst prepared from diethylaluminum chloride (Al(Et)$_2$Cl) and titanium tetrachloride (TiCl$_4$) was successful in polymerizing ethylene at normal pressure and room temperature. The polyethylene produced by this method differs significantly from that obtained by radical polymerization. The high-pressure polyethylene produced by radical polymerization is branched due to the radical transfer reactions between the polymer chains. A soft plastic with a comparatively low density is obtained, which is suitable for the production of plastic bags, for example. For this reason it is also called LD polyethylene (LD = low-density). Polymerization with the mixed catalysts developed by Ziegler and co-workers in Mühlheim produces almost unbranched linear polymer chains. The polyethylene produced in this way is a harder, semi-crystalline plastic with a higher density, from which pipes and containers can be made. This plastic is referred to as high-density (HD) PE. Figure 13.6 shows the difference between the two polymers schematically. For this achievement (polymerization of ethylene at room temperature and normal pressure) Karl Ziegler was awarded the Nobel Prize in Chemistry in 1963 together with G. Natta (stereospecific polymerization of propylene). The catalysts used are called Ziegler-Natta catalysts (ZN catalysts), mixed organometallic catalysts, or also Mülheim catalysts. They are a combination of transition metal compounds (usually halide) and main group metal alkyls (also -arylene or -hydrides) that catalytically polymerize ethene or α-olefins. The classic example is TiCl$_n$ + AlEt$_3$.

The discovery of ZN catalysts led to a series of investigations to provide insight into the mechanism of Ziegler-Natta polymerization. Figure 13.7 shows the generation of the catalyst from TiCl$_4$ and Al(Et)$_3$.

The titanium tetrachloride is reduced by the triethylaluminum in a first step and the resulting trivalent {TiCl$_3$}$_s$ precipitates as a fibrous solid. This solid is ethylated at the surface by triethylaluminum and free coordination sites are generated. These are responsible for the catalytic activity. Ziegler-Natta catalysts are heterogeneous catalysts in which the catalysis takes place on the surface of the solid.

The question now arose as to the active site and the actual mechanism of coordinative polymerization. Alternatives discussed were the formation of a dimetallic complex with alkyl/halogen bridged Ti and Al atoms or the formation of a cationic species by abstraction of a chloride ion. This issue could only be satisfactorily elucidated after studies on molecular systems. Homogeneous polymerization of ethylene succeeds with metallocenes as catalysts. The combination of dichloridobis(cyclopentadienyl)titanium (IV) ([TiCl$_2$(cp)$_2$]) and triethylaluminum as co-catalyst is a homogeneous catalyst system,

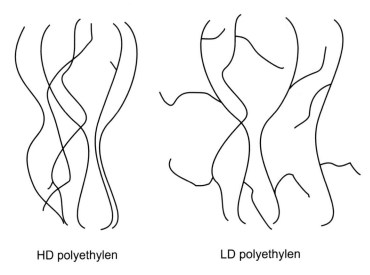

HD polyethylen   LD polyethylen

Fig. 13.6 Schematic representation of the structure of high-density (HD) polyethylene and low-density (LD) polyethylene. LD-PE obtained by radical polymerization is branched. This is the reason for the low density of this soft plastic. The HD-PE obtained by coordinative polymerization is almost unbranched, significantly harder (crystalline areas) and has a higher density

$$TiCl_4 \xrightarrow[- AlCl(Et)_2]{+ Al(Et)_3} TiCl_3(Et) \xrightarrow[- 1/2 \{C_2H_4 + C_2H_6\}]{} \{TiCl_3\}_s$$

Fig. 13.7 Generation of a Ziegler-Natta catalyst

but with low activity. This is significantly increased if MAO is used as a co-catalyst instead of pure triethylaluminum or trimethylaluminum. MAO stands for methylaluminoxane. This compound is obtained when trimethylaluminum is mixed with traces of water. Partial hydrolysis of the aluminum-carbon bond occurs and oligomeric structures are formed, which can be linear, cyclic or cage-like. Only a few of these structures are active co-catalysts, which is why MAO must be added in large excess. Its function is to methylate the catalyst and in a second step, due to the Lewis acidic nature of aluminum, to abstract a methyl anion. This produces a cationic species $[Ti(cp)_2Me]^+$, as shown in Fig. 13.8, which is discussed as the active site for the catalytic cycle. The final proof that only this cation is required for the catalytic cycle was obtained by isolating the salt $[Zr(cp)_2(Me)(THF)]^+$ BPh_4^-. This cation with the non-coordinating anion tetraphenyl borate is a polymerization catalyst even without an activator.

With knowledge of the catalytic active site, the catalytic cycle can now be formulated, which applies equally to the heterogeneous Ziegler-Natta catalysts and homogeneous metallocene catalysts. As in Ziegler's build-up reaction, a repetitive insertion of olefins into the metal-carbon bond takes place. The polymer chain grows at the metal center, which

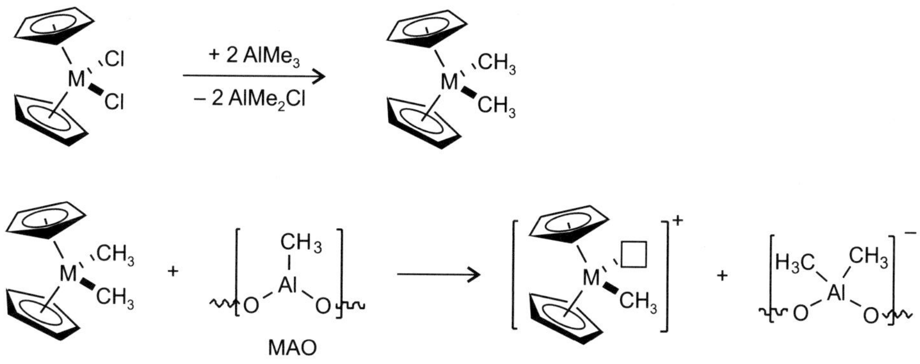

Fig. 13.8 Activation of metallocene polymerization catalysts with MAO by methylation and abstraction of a methyl anion

is why it is called a coordinative polymerization. The cycle is given in Fig. 13.9. In the first step (1→2), the ethylene coordinates to the free coordination site on the catalyst. This is followed by the insertion of the olefin into the metal-carbon bond (2→3). This insertion occurs via a cyclic transition state in which a bond is formed between the metal center and a carbon of the ethylene. Simultaneously, a bond is formed between the second ethylene carbon and the polymer chain. In this process, the bond between the polymer chain and the metal center is broken and a new free coordination site is formed. The polymer chain has changed sides from **1** to **3**. The insertion of the olefin can also be described as a *syn addition*

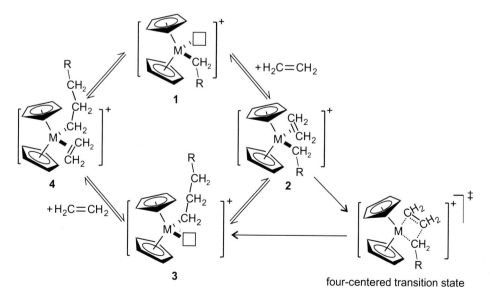

Fig. 13.9 Catalytic cycle of ethene polymerization. The four-center transition state is shown on the right. It causes the polymer chain to always migrate between two coordination sites at the metal center

of the metal-carbon bond to the C-C double bond from the olefin. Now an ethylene can bind to the free coordination site again and after insertion of the olefin into the metal-carbon bond the initial state is restored.

13.2.4 Chain Termination Reactions

In analogy to Ziegler's build-up reaction, chain termination reactions also take place during polymerization with the ZN catalysts, but only after a significantly higher number of insertions. One possible chain termination reaction – as in Ziegler's build-up reaction – is the β-H elimination. In this case, the β-H atom can be transferred either to the metal or to a monomer just coordinated to the metal. In both variants, the polymer strand formed acquires an olefinic end group and the catalyst remains active. Another chain termination reaction is homolytic cleavage of the metal-carbon bond. This variant leads to deactivation of the catalyst and polymer radicals are obtained that provide olefinic and alkyl end groups in a 1:1 ratio. Theoretically, a α-H elimination is also conceivable. Figure 13.10 shows the first two variants.

The study of metallocene catalysts for olefin polymerization has helped to find suitable models for the reaction mechanism. Another special feature of the Ziegler-Natta catalysts is their ability to stereoselectively polymerize propene. Almost exclusively isotactic polypropylene is obtained. This behavior is obtained only with the ZN catalysts, but not for the metallocenes. This raises the question of how stereoselectivity is achieved.

13.3   Polypropylene

With the Ziegler-Natta and also the metallocene catalysts, not only ethylene can be polymerized, but also longer-chain olefins such as propene (also called propylene) can be reacted. In this context, the Ziegler-Natta catalysts show a surprisingly high regio- and stereoselectivity, which was not observed with the first metallocene catalysts. Even though the metallocene catalysts are not economically interesting for the polymerization of ethylene and propylene, the studies on these systems have made an important contribution to the elucidation of the stereospecificity of the Ziegler-Natta catalysts. The findings obtained from these investigations are presented below [94].

13.3.1 Regioselectivity

In contrast to ethene, the two C atoms of the double bond of propene are not equivalent. Thus, during polymerization, the question arises between which atoms the new carbon-carbon bond is formed. If a linkage variant is preferred, then one speaks of regioselectivity. A head-tail linkage is said to occur when the bond is always formed between C1 and C2

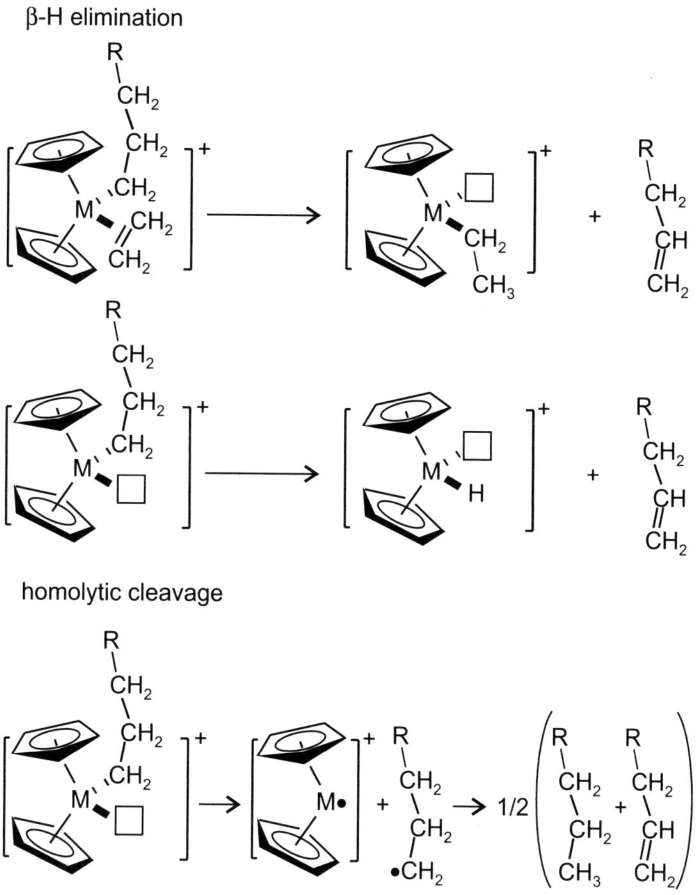

β-H elimination

homolytic cleavage

Fig. 13.10 Chain termination reactions in ethene polymerization with ZN catalysts. After both variants of *β*-H elimination, the catalyst remains active while homolytic cleavage leads to catalyst deactivation

(see Fig. 13.11). Alternatively, there could be an always alternating head-head-tail-tail linkage (C1-C1-C2-C2, see Fig. 13.11). Both variants are regioselective. If the linkage is arbitrary, no regioselectivity is observed.

Polymerization with Ziegler-Natta catalysts basically leads to a regioselective 1–2 linkage (head-tail) for propene. This circumstance can be explained by considering that it is a heterogeneous catalyst and that the methyl group (C3) has some steric demand. Figure 13.12 shows the relationship. Two variants are distinguished. In variant (a), only one [M]-C1 bond linkage takes place. Any further propene molecule is attached in such a way that the steric interaction between the methyl group and the catalyst surface is as small as possible. In variant (b), only [M]-C2 bond linkage occurs. Here, the steric claim of the methyl group of the carbon coordinated directly to the metal center causes the next propene

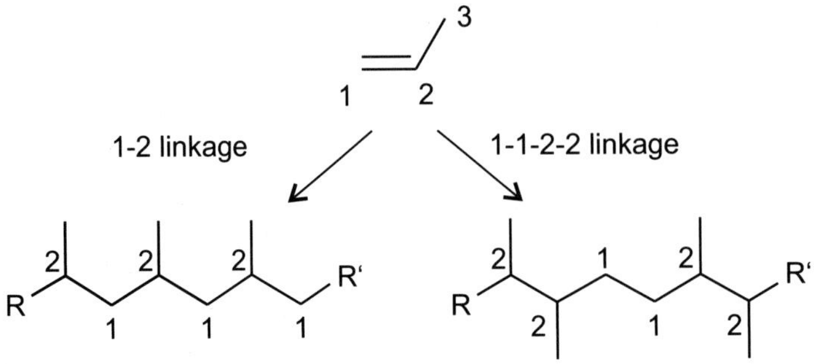

Fig. 13.11 Possibilities of regioselective polymerization of propylene. The alternating C1-C1-C2-C2 linkage cannot be realized preparatively

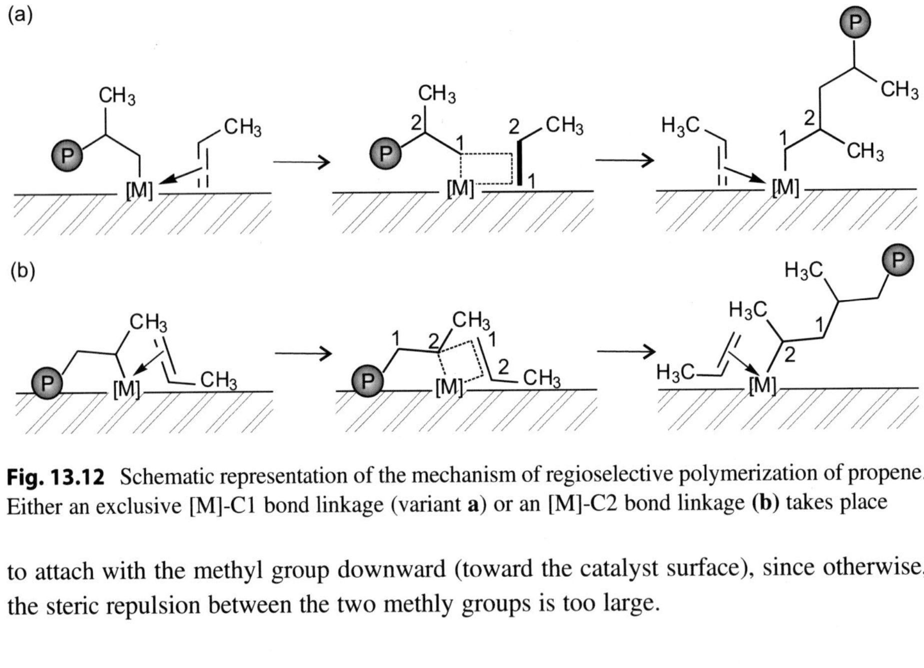

Fig. 13.12 Schematic representation of the mechanism of regioselective polymerization of propene. Either an exclusive [M]-C1 bond linkage (variant **a**) or an [M]-C2 bond linkage (**b**) takes place

to attach with the methyl group downward (toward the catalyst surface), since otherwise, the steric repulsion between the two methly groups is too large.

13.3.2 Stereoselectivity

Propene is prochiral. This means that when propene is polymerized, a chiral carbon atom is formed. A chiral carbon atom differs in all four substituents. For the carbon chains shown in Fig. 13.13, this is always true of two carbon atoms whose absolute configuration is *(R)* and *(S)*, respectively, according to the Cahn-Ingold-Prelog convention (CIP for short). One determines the absolute configuration by ordering the substituents according to their priority. In our case, the priority correlates with the number of carbon atoms in the chain

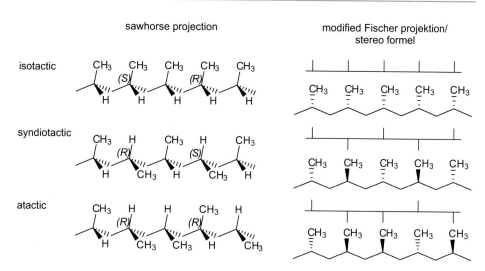

sawhorse projection

modified Fischer projektion/
stereo formel

Fig. 13.13 The regioselective polymerization of propene can yield three different stereoisomers: isotactic, syndiotactic and atactic polypropene. Atactic polymerization proceeds without stereoselectivity, whereas the other two variants are stereoselectively polymerized. Isotactic polypropene, which is obtained during polymerization with Ziegler-Natta catalysts, has the best material properties

of the substituent. The molecule is now rotated so that the substituent with the lowest priority (in our case hydrogen) points to the back (below the image plane). The remaining substituents are connected in a circular motion with decreasing priority. If this movement is done clockwise, the absolute configuration is *(R)*, counterclockwise corresponds to *(S)*.

For polypropene, the relative configuration is given instead of the absolute configuration. This is based on the first chiral atom at the left end of the chain and does not change as long as the next chiral atom has the same relative configuration. The difference can be illustrated very well by the isotactic polypropene shown in Fig. 13.13 above. The absolute configuration depends on which side of the polymer strand the chiral carbon atom is on. The relative configuration is the same for all chiral carbon atoms, with the methyl group always pointing up in the Fischer projection. Thus, the relative configuration of the three middle C atoms based on the first (left) chiral C atom is *S S S*.

The stereoselective polymerization of propene yields an isotactic or syndiotactic polymer. The first variant is obtained with the Ziegler-Natta catalysts. The metallocene catalysts presented so far polymerize without stereoselectivity. Atactic polypropene is obtained, which is the least interesting in terms of material properties.

For stereoselectivity in the polymerization of propene, it is crucial whether the catalyst coordinates on the *Re side* or the *Si side of* the propene. The relationship is shown in Fig. 13.14. Which side is *Re* or *Si* is again determined by the CIP convention. Care must be taken to ensure that the double bond has the highest priority. As illustrated in Fig. 13.14, attack on the *Re side* leads to the *S enantiomer*, while attack on the *Si side* leads to the *R enantiomer*. To obtain isotactic polypropene, the propene must always coordinate to the

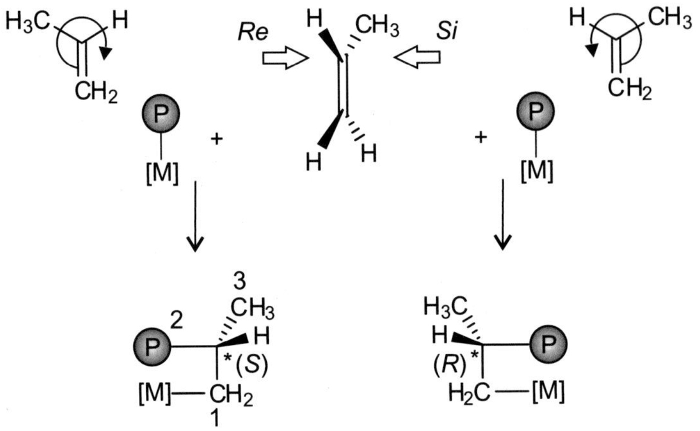

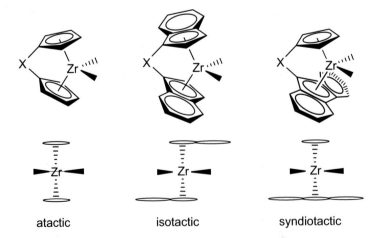

Fig. 13.14 Propene is prochiral. Depending on whether the attack of the catalyst occurs on the *Re* or *Si side*, the resulting chiral C atom has *S* or *R configuration*. When determining the priority according to the CIP convention, it must be ensured that the double bond has the highest priority in the case of the prochiral propene and the side chain with catalyst has the highest priority in the case of the product

transition metal with the same side, whereas for syndiotactic polypropene the sides must alternate.

Stereoselective Polypropylene with Metallocene Catalysts

While the Ziegler-Natta catalysts supplied stereospecific isotactic polypropylene right from the start, this was only achieved with the metallocene catalysts with the advent of ansa metallocenes. Ansa stands for handle and means that the two cyclopentadienyl rings are connected by a "handle". Three examples of such precatalysts are given in Fig. 13.15.

atactic   isotactic   syndiotactic

Fig. 13.15 Stereospecific polymerization of propene with metallocene catalysts. Depending on the subsitution pattern on the cyclopentadienly ring, atactic, isotactic or syndiotactic polypropene is obtained. The bridging group X is e.g. SiMe$_2$

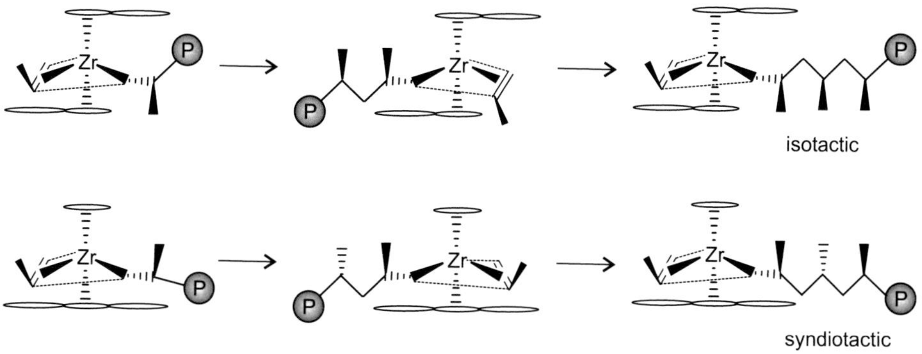

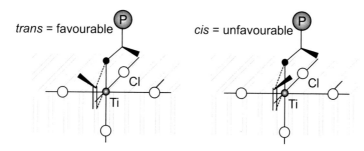

isotactic

syndiotactic

Fig. 13.16 Schematic representation of the mechanism of the stereospecific polymerization of propene with metallocene catalysts

trans = favourable   *cis* = unfavourable

Fig. 13.17 Schematic representation of the mechanism of the stereospecific polymerization of propene with Ziegler-Natta catalysts

Depending on the substitution pattern on the cp rings, the stereoselectivity in the polymerization of propene can be adjusted.

Figure 13.16 gives the mechanism for the preparation of isotactic and syndiotactic polypropene. Because of the steric repulsion of the substituent on the cyclopentadienyl ring, there is always only one variant of how the entering propene molecule can coordinate to the catalyst. These findings can now be applied to the Ziegler-Natta catalysts and lead to the suggestion given in Fig. 13.17 for the preferred coordination of propene to the active site on the catalyst surface.

13.4 Photocatalysis

13.4.1 Basics

In the section on luminescence in complexes, we have already considered in detail the electronic states of complexes excited with light. In addition to the non-radiative deactivation or emission of light, there is a third possibility for which this state can be used: for the realization of chemical reactions that would not take place without the irradiation of light. Before we look at this in detail, let us ask ourselves what the difference is between photocatalysis and a photochemical reaction.

In a photochemical reaction, light is involved in the reaction in the form of photons. The photons are absorbed by one of the reactants before the reaction, i.e. consumed during the reaction. This causal relationship between absorption and reaction was already formulated by Grotthuss (1817) and a second time by Draper (1842). According to the definition of catalysis, photons themselves cannot have a catalytic effect, since a catalyst is available again, unconsumed, after the reaction. The light in this case is not a catalyst but a starting material. As an example, let us consider the following simple redox reaction between an acceptor molecule A and a donor molecule D in solution:

$$A + D + h\nu \rightarrow A^* + D \rightarrow A^- + D^+$$

Figure 13.18 shows the energy diagrams associated with the reaction. The activation energy for the redox reaction between A and B is very high in the dark, so under these conditions the reaction does not proceed. When A absorbs light of the appropriate wavelength, we get the electronically excited state A*. Here an electron has been excited into a higher energy, often antibonding, molecular orbital (the LUMO). In addition, a positive "hole" remains in the previously fully occupied HOMO. In the electronically excited state, compound A* can now more easily donate the "energy-rich" electron and thus acts as a reducing agent. At the same time, the "hole" can accept an electron. A* can therefore also act as an electron acceptor, as an oxidizing agent. The two do not contradict

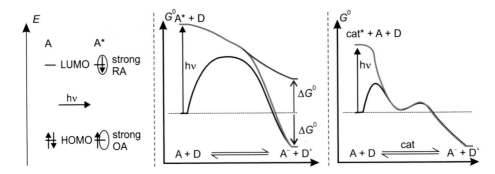

Fig. 13.18 Left: when an electron is excited with light from the highest occupied molecular orbital (HOMO) to the lowest unoccupied molecular orbital (LUMO), this electronically excited state has both a stronger oxidation capacity and reduction capacity. Middle: Two possible energy profiles of such a photochemical reaction. Starting from the excited state, a different reaction path is taken than in the thermal reaction. In some photochemical reactions the energy of the products is higher than that of the reactants (red curve). This is called photochemical energy storage, an example being photosynthesis. Right: Comparison of the energy profile of a catalysed and a photocatalysed reaction

each other, A* is a stronger oxidizing and reducing agent at the same time compared to A. In our example reaction, A acts as an acceptor and is reduced.

If the free enthalpy G^0 of the products A^- and D^+ is now higher than that of the reactants A and D, such reactions are also referred to as photochemical energy storage (see Fig. 13.18). This condition is met, for example, in photosynthesis, where the products (oxygen and sugars) are more energetic than the reactants (CO_2 and H_2O). Another example would be the splitting of water to hydrogen and oxygen.

The term photocatalysis is usually used in connection with a photocatalyst. In this context, the photocatalyst, PC, is not light, but a substance that generates the state {D-PC-A}* with induced activity in the presence of light. Thus, catalytically active is the excited state of the photocatalyst. This fundamentally distinguishes photocatalysis from thermal catalysts with permanent activity. The reactants do not normally absorb any light:

$$A + D \xrightarrow[PC]{h\nu} A^- + D^+$$

Photocatalysts can be molecules or complexes in which the excited state is long-lived enough to participate in subsequent reactions. In most cases, these photocatalysts are present dissolved with the starting materials in the reaction mixture, i.e. homogeneous catalysis takes place. However, semiconductors such as TiO_2, ZnO or CdS can also be used as photocatalysts for heterogeneous photocatalysis. Figure 13.19 shows a comparison of the two variants. Largely analogous to the excitation of an electron from a HOMO into a LUMO in molecular systems, in semiconductors the electron is excited from the valence band (VB) into the conduction band (CB). In this way, electron/hole (or electron/defect

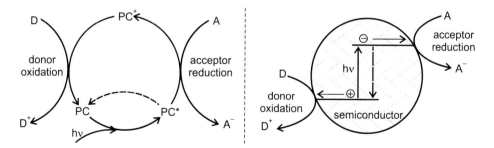

Fig. 13.19 Left: Schematic representation of a redox reaction with a molecular photocatalyst (PC) in homogeneous phase. The quenching from the excited state PC* occurs with electron donation to A. However, the complementary case, the quenching of PC* with electron acceptance from D, is equally possible. Both ways are possible. Right: Schematic representation of a semiconductor particle as a photocatalyst for redox reactions. The electron and the hole are shown here as positive and negative charge, which migrates to the particle surface and is available there for subsequent reactions. In both variants, not every absorbed photon leads to a subsequent chemical reaction. The reason for this is the process shown as a dashed line, where the molecule or semiconductor falls back to the electronic ground state without participating in a reaction. In both cases the PC is unchanged after reaction

electron) pair formation also occurs in the electronically excited state, and the semiconductor also acts as a good oxidizing and reducing agent. In order to be available for reaction with solute and/or adsorbed partners, the electrons and the defect electrons must migrate to the surface of the semiconductor, a process also known as exiton migration. Spatially separated charges are created. In the case of molecular systems, we have already learned about the possibility of radiationless deactivation for the return of the system to the ground state in the chapter on luminescence. This process always occurs as an alternative to the desired subsequent chemical reaction. In semiconductors, this process also occurs and is called recombination of the electron/hole pair. This deactivation of the electronically excited state means that not every photon irradiated is also used for a chemical reaction. The efficiency of a photocatalytic reaction can be indicated by the photocatalytic quantum yield or quantum efficiency.

Finally, we ask ourselves what the difference is between a photocatalyst and a photosensitizer (PS). For molecular systems, this question can be answered very well. The system shown in Fig. 13.19 on the left is a photocatalyst that participates directly in the redox steps of the chemical reaction. In the excited state, PC* is first oxidized by the acceptor molecule A to PC$^+$ (alternatively, PC* can be reduced by the donor to PC$^-$), and then in a further step is reduced again to PC by the donor molecule D. Absorption of a quantum of light again starts the next cycle. Ultimately, PC* acts as a redox mediator in terms of a pair of (photo)electrolytic one-electron processes. In a photosensitized reaction, the photosensitizer PS is transferred to the electronically excited state PS* by light excitation. This excited state returns to the ground state, transferring energy to an antenna molecule, which then catalyzes the subsequent reactions. A nice example for this from nature is the light-collecting complex in photosynthesis, which absorbs sunlight and transfers the energy to the reaction center of the photosynthesis light reaction, where the actual catalysis takes place.

For semiconductors in heterogeneous photocatalysis, this discussion catalysis vs. sensitization is much more controversial, since it is difficult to show that the semiconductor is temporarily chemically changed by oxidation or reduction. One could also speak of a photoreaction sensitized by semiconductor. It becomes even more complex when not only a semiconductor, but a combination of semiconductor and noble metal (e.g. gold particles on TiO$_2$) is used as a heterogeneous catalyst. In the example given, one could assume that TiO$_2$ is responsible for the absorption of light, while the actual catalysis takes place on the gold particle. Thus, TiO$_2$ would be the photosensitizer, or absorber. For other composite materials, for example the combination of two semiconductors (TiO$_2$ and CdS), the assignment of the individual functions becomes much more difficult. In the context of this book chapter, we refer to both the molecular systems and the semiconductors that function as in Fig. 13.19 as photocatalysts. In the case of composite materials, the term often refers to the complete material system.

13.4.2 Photosynthesis

Photosynthesis is the best known example of a photocatalytic reaction optimized by nature. In the following, we will consider only the light reaction that takes place in the chloroplasts of all green plants. The whole process is shown schematically in a highly simplified form in Fig. 13.20. Two photosystems are involved in the light reaction, designated P700 (photosystem, PS I) and P680 (PS II) according to the wavelength absorbed. In PS II, water oxidation takes place at the oxygen-evolving complex (OEC), whereas in PS I, NADPH (NADP = nicotinic acid amide adenine dinucleotide phosphate, hydride ion-transferring coenzyme) is formed. The two synchronized excitations must be coupled together because the redox process $NADP^+$ to NADPH requires a redox potential, which can only be provided by PS I. In the biological system, the redox mediators R shown in the diagram are complex electron transfer chains that enable rapid charge separation via one or more steps. Examples of complexes used in such electron transfer chains were discussed in the

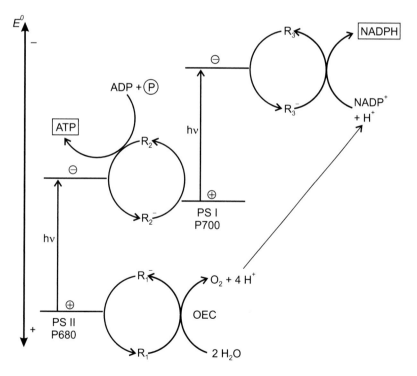

Fig. 13.20 Highly simplified schematic of the mechanism of photosynthesis. The positive holes and excited electrons of the two photosystems (P680 and P700) are brought to the catalytic active sites via complex electron transfer chains, abbreviated here as redox mediators R_n. The holes formed by the P680 are used to generate oxygen at the oxygen-volving complex (OEC). The excited electrons from the P700 are used to store hydrogen in the form of NADPH. During the transition from PS II to PS I, the energy carrier ATP is also generated

chapter on redox reactions in complexes (blue copper proteins). The redox mediator R_2 is the link between the two photosystems. The coupling releases some energy which is used for the synthesis of the biological energy carrier ATP (adenosine triphosphate). The products ATP and NADPH are important starting materials for the dark reaction, in which carbon dioxide and water are converted to carbohydrates under energy consumption (endergonic), while oxygen is released to the environment as a by-product.

We note that photosynthesis is a very complex process in which many individual steps are intertwined and precisely coordinated (spatial orientation, redox potentials). Nevertheless, the efficiency of photosynthesis is very low, 0.1–1%, and most of the adsorbed light is not converted into chemical energy. Nature solves this problem by the mass of leaves that plants form. This works because all processes rely on readily available elements, including catalytic active sites. In order to absorb as much light energy as possible and to cover a broad wavelength range, additional photosensitizers (chlorophylls, carotenoids) are used in the light-collecting complexes.

The Oxygen-Evolving Complex (OEC)

Before we turn to examples of artificial photosynthesis, the photocatalytic water splitting, we look at the OEC from PS II in a little more detail. The process that takes place here, water oxidation, is also a great challenge for photocatalytic water splitting because it is a four-electron process in which the electron transfer is additionally coupled to a proton transfer.

$$2H_2O \xrightarrow{h\nu} O_2 + 4H^+ + 4e^-$$

The whole process is catalyzed by readily available elements (manganese and calcium). Accordingly, the interest in understanding the whole process in detail is high, in order to transfer the knowledge gained to artificial systems. Figure 13.21 shows the catalytic cycle and the structure of the Mn_4Ca cluster in the OEC.

PS II is located in the thylakoid membrane of chloroplasts. Long before the structure of the OEC was elucidated, Kok proposed a five-state model for water oxidation in 1970 based on a flash experiment. Starting from the maximally reduced state S_0, the cluster is oxidized stepwise in four successive $1e^-/H^+$ steps until oxygen release then occurs at the most highly oxidized state S_4. A very well elucidated structure from the OEC exists in the S_1 state, which is also present in the dark as the resting state and is the ground state of the system, see Fig. 13.21. Here, two manganese(III) and two manganese(IV) ions are each present in an octahedral coordination environment. The two oxidation states can be readily assigned due to the strong Jahn-Teller distortion in the high-spin $3d^4$ ion Mn^{3+} (the preferential coordination of weak-field ligands such as carboxylate groups, water, and hydroxide stabilizes the high-spin state), compared to the nearly undistorted $3d^3$ ion Mn^{4+}. The Jahn-Teller effect has already been discussed for the blue copper proteins using the $3d^9$ copper(II) ion as an example (Chap. 7), where an E state is present. The manganese(III) ion

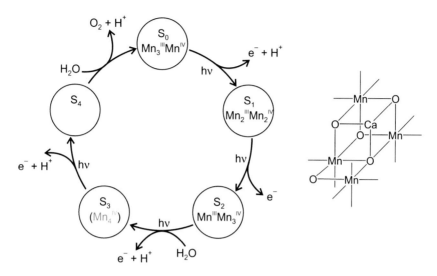

Fig. 13.21 Schematic representation of the catalytic cycle in the OEC and structure of the Mn$_4$Ca cluster in the OEC where oxygen evolution occurs. The assignment of oxidation states at S$_3$ and S$_4$ for the Magnan centers is not yet certain. The further bonds indicated are saturated by carboxylate groups of amino acids, a histidine residue and by water/hydroxido ligands. The increasingly positive oxidation states during the catalytic cycle are stabilized by the deprotonation of ligands (increasing negative charge). The excitation steps painted in the cycle (hν) do not occur in the manganese cluster, as shown in Fig. 13.20. The catalytic agent here is a tyrosyl radical. This is generated by scavenging the oxidized state of PSII from tyrosinate. The mediator R$_1$ in Fig. 13.20 is therefore a tyrosyl radical

is the second biologically relevant ion with a pronounced tendency to Jahn-Teller distortion, again due to a twofold degeneracy of the ground state.

The structures of the states S$_0$ and S$_2$ are also still largely certain, the structure and oxidation states of S$_3$ are still discussed in the literature while the structure of S$_4$, i.e. the most highly oxidized state where oxygen evolution takes place, is not yet certain. Accordingly, the questions of when (S$_3$ and S$_4$ or only S$_4$), where (two Mn or one Mn and the Ca) and how the two water molecules bind and how the oxygen is formed are still unresolved and the subject of current research.

13.4.3 Photocatalytic Water Splitting

Photocatalytic water splitting is virtually the artificial analogue of photosynthesis and one of the most studied reactions for storing solar energy in fuel. Hydrogen as a green energy carrier from water splitting is a highly attractive basic chemical. Solar hydrogen can be used to power fuel cells, it can be used as a substitute for hydrogen from fossil fuels for a variety of large-scale chemical conversions (ammonia production, reduction of iron oxides with H$_2$ as an alternative to the blast furnace process). Thus, the goal of research on photocatalytic water splitting is not only to model PS II, but primarily to develop a technically effective catalysis for energy storage. Considering the complex interplay of several components in photosynthesis, it is not surprising that photocatalytic water splitting also only works with coupled systems. In the following, two examples are briefly presented.

One of the first examples of complete water splitting is given in Fig. 13.22 above. The photocatalyst is $[Ru(bipy)_3]^{2+}$, whose photochemistry (and physics) we have already discussed in detail in the luminescence of complexes (Chap. 11). Methyl viologen serves as a redox mediator for rapid charge separation, similar to electron transfer chains in the biological system; the chemical formula is given in the figure. For water oxidation, ruthenium(IV) oxide is used as catalyst and colloidally distributed platinum nanoparticles are used for hydrogen evolution. The efficiency for complete water splitting (water photolysis) in this reaction is $10^{-4}\%$, again much lower than photosynthesis. [98] When a sacrificial donor such as triethanolamine or EDTA takes the place of water and is irreversibly oxidized, the efficiency increases to 20% (quantum yield, molecules converted per quanta of light absorbed) for hydrogen evolution. [99] A high number of cycles, which is very relevant for practical applications, has not yet been achieved in either example. It requires a high chemical and photochemical stability of all catalytic components involved.

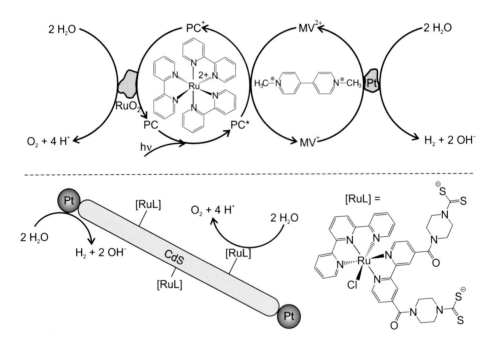

Fig. 13.22 Top: Example of one of the first systems for complete photocatalytic water splitting, consisting of several individual components. Bottom: Actual example of a composite material consisting of CdS nanorods decorated with platinum nanoparticles and a ruthenium complex, which can be used to realize complete water splitting with visible light

In a very recent example, platinum nanoparticles and molecular ruthenium catalysts are attached to CdS nanorods. In this way, all the components involved are fixed in spatial proximity to each other. As a semiconductor, the CdS serves as an absorber of light in the visible range and causes primary charge separation. The Pt nanoparticles at the ends of the rods catalyze hydrogen evolution (reduction) with a quantum efficiency of up to 4.9%, analogous to the previous example. Oxygen evolution is catalyzed by a molecular catalyst, the complex [Ru(tpy)(bpy)Cl], which was fixed to the surfaces of the nanorods via suitable anchor groups (see Fig. 13.22 below). In this substep, the quantum efficiency is significantly lower, up to 0.27%, and we have already discussed the reasons for this in the OEC. [100] The example shows that by the specific design of composite materials, artificial photosynthesis with quite comparable, or even better, efficiency is possible.

13.5 Questions

1. Divide the cycle of polymerization catalysis into the elementary reactions of organometallic chemistry!
2. Determine the electron number of the complexes at all catalytic intermediates.
3. Why does Ziegler's build-up reaction yield only short olefins in the presence of nickel and long polymers in the presence of titanium(IV)?
4. Photosynthesis is often used as an example of successful photocatalysis.
   - Briefly explain the general mode of action of a photocatalyst.
   - What are the special features of photosynthesis? In doing so, elaborate on the OEC and the function of PS1 and PS2.
   - Briefly discuss energy efficiency.

References

1. Holleman, A. F., Wiberg, E., & Wiberg, N. (2007). *Lehrbuch der anorganischen Chemie* (102. Aufl.). de Gruyter.
2. Kober, F. (1979). *Grundlagen der Komplexchemie* (1. Aufl.). O. Salle.
3. Gade, L. H., & Lewis, J. (1998). *Koordinationschemie*. Wiley-VCH.
4. Kauffman, G. B. (1994). *Coordination Chemistry: A century of progress*. American Chemical Society.
5. Gade, L. H. (2002). *Chemie in unserer Zeit, 36*, 168–175.
6. Woodward, J. (1724). *Philosophical Transactions of the Royal Society of London, 33*, 15–17.
7. Brown, J. (1724). *Philosophical Transactions of the Royal Society of London, 33*, 17–24.
8. Ozeki, T., Matsumoto, K., & Hikime, S. (1984). *Analytical Chemistry, 56*, 2819–2822.
9. Izatt, R. M., Watt, G. D., Bartholomew, C. H., & Christensen, J. J. (1970). *Inorganic Chemistry, 9*, 2019–2021.
10. Heyn, B., Hipler, B., Kreisel, G., Schreer, H., & Walther D. (1990). *Anorganische Synthesechemie: Ein integriertes Praktikum* (2. Aufl.). Springer.
11. Werner, A. (1893). *Zeitschrift für anorganische und allgemeine Chemie, 3*, 267–330.
12. Neil, G. (2005). *Connelly Nomenclature of inorganic chemistry: IUPAC recommendations 2005*. RSC Publishing.
13. Weber, B., Betz, R., Bauer, W., & Schlamp, S. (2011). *Zeitschrift für anorganische und allgemeine Chemie, 637*, 102–107.
14. Florian Kraus, unpublished results.
15. Elschenbroich, C. (2008). *Organometallchemie*, (6. Aufl.). Teubner.
16. Zeise, W. C. (1831). *Annalen der Physik und Chemie, 97*, 497–541.
17. Hieber, W., & Leutert, F. (1931). *Naturwissenschaften, 19*, 360–361.
18. David Manthey., Orbital Viewer, Version 1.04. http://www.orbitals.com/orb.
19. Hauser, A. (1991). *The Journal of Chemical Physics, 94*, 2741.
20. Weber, B., Käpplinger, I., Görls, H., & Jäger, E.-G. (2005). *European Journal of Inorganic Chemistry, 2005*, 2794–2811.
21. Lind, M. D., Hamor, M. J., Hamor, T. A., & Hoard, J. L. (1964). *Inorganic Chemistry, 3*, 34–43.
22. Botta, M., Lee, G. H., Chang, Y., Kim, T.-J. (2012). *European Journal of Inorganic Chemistry*. 1924–1933.
23. Taube, H. (1984). *Angewandte Chemie International Edition in English, 23*, 329–339.
24. Marcus, R. A. (1956). *The Journal of Chemical Physics, 24*, 966.
25. Enemark, J. H., & Feltham, R. D. (1974). *Coordination Chemistry Reviews, 13*, 339–406.

B. Weber, *Coordination Chemistry*,
https://doi.org/10.1007/978-3-662-66441-4

26. Wanat, A., Schneppensieper, T., Stochel, G., van Eldik, R., Bill, E., & Wieghardt, K. (2002). *Inorganic Chemistry, 41*, 4–10.
27. Cram, D. J. (1988). *Angewandte Chemie International Edition, 27*, 1009–1020.
28. Lehn, J.-M. (1988). *Angewandte Chemie International Edition, 27*, 89–112.
29. Pedersen, C. J. (1988). *Angewandte Chemie International Edition, 27*, 1021–1027.
30. Swiegers, G. F., & Malefetse, T. J. (2000). *Chemical Reviews, 100*, 3483–3538.
31. Batten, S. R., Champness, N. R., Chen, X.-M., Garcia-Martinez, J., Kitagawa, S., Öhrström, L., O'Keeffe, M., Paik Suh, M., & Reedijk, J. (2013). *Pure and Applied Chemistry, 85*, 1715–1724.
32. Kaskel, S. (2005). *Nachrichten aus der Chemie, 2005*, 394–399.
33. Eddaoudi, M., Moler, D. B., Li, H., Chen, B., Reineke, T. M., O'Keeffe, M., & Yaghi, O. M. (2001). *Accounts of Chemical Research, 34*, 319–330.
34. Chen, B., Fronczek, F. R., Courtney, B. H., & Zapata, F. (2006). *Crystal Growth & Design, 6*, 825–828.
35. Czaja, A. U., Trukhan, N., & Müller, U. (2009). *Chemical Society Reviews, 38*, 1284.
36. Murray, L. J., Dincă, M., & Long, J. R. (2009). *Chemical Society Reviews, 38*, 1294.
37. Sun, D., Ma, S., Ke, Y., Collins, D. J., & Zhou, H.-C. (2006). *Journal of the American Chemical Society, 128*, 3896–3897.
38. Nguyen, T. (2005). *Science, 310*, 844–847.
39. Brynda, M., Gagliardi, L., Widmark, P.-O., Power, P. P., & Roos, B. O. (2006). *Angewandte Chemie International Edition, 45*, 3804–3807.
40. Wagner, F. R., Noor, A., & Kempe, R. (2009). *Nature Chemistry, 1*, 529–536.
41. Hoffmann, R. (1982). *Angewandte Chemie, 94*, 725–739.
42. Gordon, F., & Stone, A. (1984). *Angewandte Chemie, 96*, 85–96.
43. Hatscher, S., Schilder, H., Lueken, H., & Urland, W. (2005). *Pure and Applied Chemistry, 77*, 497–511.
44. Kahn, O. (1993). *Molecular Magnetism*. VCH.
45. Mabbs, F. E., & Machin, D. J. (1973). *Magnetism and transition metal complexes*. Chapman and Hall.
46. Heiko Lueken Magnetochemie. (1999). *Eine Einführung in Theorie und Anwendung*. Teubner Studienbücher Chemie Teubner.
47. Landrum, G. A., & Dronskowski, R. (1999). *Angewandte Chemie, 111*, 1481–1485.
48. Kahn, O., Galy, J., Journaux, Y., Jaud, J., & Morgenstern-Badarau, I. (1982). *Journal of the American Chemical Society, 104*, 2165–2176.
49. Anderson, P. W. (1959). *Physical Review, 115*, 2–13.
50. Longuet-Higgins, H. C. (1950). *The Journal of Chemical Physics, 18*, 265.
51. Glaser, T., Heidemeier, M., Grimme, S., & Bill, E. (2004). *Inorganic Chemistry, 43*, 5192–5194.
52. Anderson, P. W., & Hasegawa, H. (1955). *Physical Review, 100*, 675–681.
53. Gütlich, P., & Goodwin, H. A. (Hrsg.). (2004). *Spin crossover in transition metal compounds*; Bd. 233–235 of *Topics in current chemistry*. Springer.
54. Halcrow, M. A. (Ed.). (2013). *Spin crossover materials: Properties and applications*. Wiley-Blackwell.
55. Cambi, L., & Szegö, L. (1933). *Berichte der deutschen chemischen Gesellschaft (A and B Series), 66*, 656–661.
56. Baker, W. A., & Bobonich, H. M. (1964). *Inorganic Chemistry, 3*, 1184–1188.
57. Decurtins, S., Gütlich, P., Köhler, C. P., Spiering, H., & Hauser, A. (1984). *Chemical Physics Letters, 105*, 1–4.
58. Decurtins, S., Gutlich, P., Hasselbach, K. M., Hauser, A., & Spiering, H. (1985). *Inorganic Chemistry, 24*, 2174–2178.

59. Köhler, C. P., Jakobi, R., Meissner, E., Wiehl, L., Spiering, H., & Gütlich, P. (1990). *Journal of Physics and Chemistry of Solids, 51,* 239–247.
60. Niel, V., Carmen Muñoz, M., Gaspar, A. B., Galet, A., Levchenko, G., & Real, J. A. (2002). *Chemistry – A European Journal, 8,* 2446–2453.
61. Hauser, A. (1991). *Coordination Chemistry Reviews, 111,* 275–290.
62. Hauser, A., Enachescu, C., Daku, M. L., Vargas, A., & Amstutz, N. (2006). *Coordination Chemistry Reviews, 250,* 1642–1652.
63. Létard, J.-F. (2006). *Journal of Materials Chemistry, 16,* 2550.
64. Weber, B. (2013). In M. A. Halcrow (Ed.), *Novel Mononuclear Spin Crossover Complexes.* Wiley-Blackwell.
65. Bonhommeau, S., Molnár, G., Galet, A., Zwick, A., Real, J.-A., McGarvey, J. J., & Bousseksou, A. (2005). *Angewandte Chemie, 117,* 4137–4141.
66. Vankó, G., Renz, F., Molnár, G., Neisius, T., & Kárpáti, S. (2007). *Angewandte Chemie International Edition, 46,* 5306–5309.
67. Gütlich, P., Hauser, A., & Spiering, H. (1994). *Angewandte Chemie, 106,* 2109–2141.
68. König, E. (1991). *Nature and dynamics of the spin-state interconversion in metal complexes,* Bd. 76. Springer.
69. Real, J. A., Gaspar, A. B., Niel, V., & Carmen Muñoz, M. (2003). *Coordination Chemistry Reviews, 236,* 121–141.
70. Weber, B. (2009). *Coordination Chemistry Reviews, 253,* 2432–2449.
71. Rotaru, A., Carmona, A., Combaud, F., Linares, J., Stancu, A., & Nasser, J. (2009). *Polyhedron, 28,* 1684–1687.
72. Weber, B., Bauer, W., & Obel, J. (2008). *Angewandte Chemie, 120,* 10252–10255.
73. Létard, J.-F., Guionneau, P., Codjovi, E., Lavastre, O., Bravic, G., Chasseau, D., & Kahn, O. (1997). *Journal of the American Chemical Society, 119,* 10861–10862.
74. Zhong, Z. J., Tao, J.-Q., Yu, Z., Dun, C.-Y., Liu, Y.-J., & You, X.-Z. (1998). *Journal of the Chemical Society, Dalton Transactions,* 327–328.
75. Förster, C., & Heinze, K. (2020). *Chemical Society Reviews, 49,* 1057–1070.
76. Valeur, B., & Berberan-Santos, M. N. (2013). *Molecular Fluorescence: Principles and applications,* (2. Aufl.). Wiley-VCH.
77. Balzani, V., Moggi, L., Manfrin, M. F., Bolletta, F., & Laurence, G. S. (1975). *Coordination Chemistry Reviews, 15,* 321–433.
78. Balzani, V., Ceroni, P., & Juris, A. (2014). *Photochemistry and photophysics: Concepts, Research.* Wiley-VCH.
79. Balzani, V., Bergamini, G., Campagna, S., & Puntoriero, F. (2007). *Photochemistry and Photophysics of Coordination Compounds: Overview and General Concepts.* in *Topics in Current Chemistry* Bd. 280, Springer.
80. Mustroph, H., & Ernst, S. (2011). *Chemie in Unserer Zeit, 45,* 256–269.
81. Förster, C., & Heinze, K. (2020). *Journal of Chemical Education, 97,* 1644–1649.
82. Otto, S., Grabolle, M., Förster, C., Kreitner, C., Resch-Genger, U., & Heinze, K. (2015). *Angewandte Chemie, 127,* 11735–11739.
83. Mauro, M., Aliprandi, A., Septiadi, D., Kehra, N. S., & De Cola, L. (2014). *Chemical Society Reviews, 43,* 4144–4166.
84. Puttock, E. V., Walden, M. T., & Gareth Williams, J. A. (2018). *Coordination Chemistry Reviews, 367,* 127–162.
85. Aliprandi, A., & Mauro, M. (2016). *Luisa De Cola Nature Chemistry, 8,* 10–15.
86. Kaim, W., & Schwederski, B. (2010). *Bioanorganische Chemie: Zur Funktion chemischer Elemente in Lebensprozessen,* (4. Aufl.). STUDIUM Vieweg+Teubner.
87. Herres-Pawlis, S., & Klüfers, P. (2017). *Bioanorganishe Chemie: Metalloproteine.* Wiley-VCH.

88. Suslick, K. S., & Reinert, T. J. (1985). *Journal of Chemical Education, 62*, 974.
89. Collman, J. P., Boulatov, R., Sunderland, C. J., & Lei, F. (2004). *Chemical Reviews, 104*, 561–588.
90. Männel-Croise, C., Probst, B., & Zelder, F. (2009). *Analytical Chemistry, 81*, 9493–9498.
91. Pattenden, G. (1988). *Chemical Society Reviews, 17*, 361.
92. König, B. (2013). *Chemical Photocatalysis*. De Gruyter.
93. Fischer, K., Jonas, K., Misbach, P., Stabba, R., & Wilke, G. (1973). *Angewandte Chemie, 85*, 1001–1012.
94. Brintzinger, H.-H., Fischer, D., Mülhaupt, R., Rieger, B., & Waymouth, R. (1995). *Angewandte Chemie, 107*, 1255–1283.
95. Wurzenberger, X., Piotrowski, H., & Klüfers, P. (2011). *Angewandte Chemie, 123*, 5078–5082.
96. Monsch, G., & Klüfers, P. (2019). *Angewandte Chemie, 131*, 8654–8659.
97. Ampßler, T., Monsch, G., Popp, J., Riggenmann, T., Salvador, P., Schröder, D., & Klüfers, P. (2020). *Angewandte Chemie, 132*, 12480–12485.
98. Kiwi, J., & Grätzel, M. (1979). *Nature, 281*, 657–658.
99. Kalyanasundaram, K., Grätzel, M., & Pelizzetti, E. (1986). *Coordination Chemistry Reviews, 69*, 57–125.
100. Wolff, C. M., Frischmann, P. D., Schulze, M., Bohn, B. J., Wein, R., Livadas, P., Carlson, M. T., Jäckel, F., Feldmann, J., Würthner, F., & Stolarczyk, J. K. (2018). *Nature Energy, 3*, 862–869.